Defect Prevention
Solutions Manual

QUALITY AND RELIABILITY

A Series Edited by

Edward G. Schilling

Center for Quality and Applied Statistics
Rochester Institute of Technology
Rochester, New York

Defect Prevention
Use of Simple Statistical Tools
Solutions Manual

Victor E. Kane

Ford Motor Company
Livonia, Michigan

Sponsored by the American Society for Quality Control

Marcel Dekker, Inc.
ASQC Quality Press

New York and Basel
Milwaukee

ISBN 0-8247-8163-5

This book is printed on acid-free paper.

MARCEL DEKKER, INC.
270 Madison Avenue, New York, New York 10016

Current printing (last digit):
10 9 8 7 6 5 4 3 2 1

PRINTED IN THE UNITED STATES OF AMERICA

Contents

Solutions to the Problems in Chapter 2

2.1 The two paths show how a process flow diagram can be useful in assessing sources of variability:

Sources of variability	Problem identified at location		
	Output A	Output D	Output B
Input	A	B, C	B
Machine	A	B, C	B
Spindles	(all eight)	(all eight)	(all four on machine B)

The problem shows how points of control are influenced by different sources of variability.

2.2 The following are the potential paths:

	Machine					
Suppliers	A	B	C	D	E	F
2	$\times$ 2 $\times$	(3	+ 3 +	3)	$\times$ (4 +	4) = 288

2.3 The following are the potential paths:

	Operation							
Supplier	20	30	40	50	60	65	80	90
1	$\times$ 4 $\times$	24 $\times$	16 $\times$	6 $\times$	12 $\times$	1 $\times$	1 $\times$	2 = 221,184

2.4 Target path flows can be used as follows:

Original paths:
Supplier	A, B, C, D	E
2 $\times$	8	$\times$ 3 = 48

Targeted paths:
1 $\times$	4	$\times$ 3 = 12

1

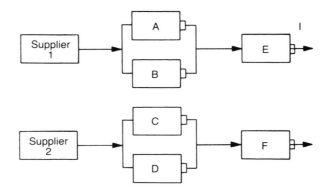

2.5 The six measurements are plotted for each comparison.

(a) Gage 1 more accurate, less repeatable than gage 2:

```
Gage 1:          x      x  x    x           x      x
            8    9    10    11    12    13    14    15

Gage 2:        xx    x  x   x x
            8    9    10    11    12    13    14    15
```

(b) Gage 1 more repeatable than gage 2. Both gages have about the same accuracy:

```
Gage 1:                   x    x xx    x     x
            8    9    10    11    12    13    14    15

Gage 2:        x      x      x           x      x x
            8    9    10    11    12    13    14    15
```

(c) Gage 1 less accurate than gage 2. Both gages have about the same repeatability:

```
Gage 1:     x x    x x x  x
            8    9    10    11    12    13    14    15

Gage 2:                                x    x x x   xx
            8    9    10    11    12    13    14    15
```

Solutions to the Problems in Chapter 3

3.1 Answers are listed in order $\bar{X}$, $\tilde{X}$, s, and R: (a) 4.3, 4, 2.52, 5; (b) 3.98, 2.50, 3.93, 8.5; (c) 6.86, 7.1, 2.36, 6.1; (d) 2.33, 3, 3.06, 6; (e) -3.2, -3, 5.54, 14; (f) 3.58, 4.61, 4.27, 8.36; (g) 8.6, 9, 2.1, 5; (h) 1.03, 0, 4.74, 9.3.

3.2 (a) 425, 453, 552, 433, 449.5; (b) -30, 52, 13, -36.6, 23

3.3 .0024, .0178, .0015, and .0007.

3.4 (a) 29.4201, 29.42155, 29.4200, 29.41965, 29.4184; (b) 29.421, 29.419; (c) 6.2, -9.0, 30, -10, 5.7; (d) actual dimension value $= 29.420 + (.0001 \times$ gage reading).

3.5 (a) The computed volumes are $\bar{X} = 37$, $\tilde{X} = 38$, $s = 4.92$, and $R = 16$. The process spread is $6s = 29.5$ or $6R/d_2 = 96/3.81 = 25.2$. The expected high and low values in the process are:

$$\begin{aligned}
\text{UPL} &= \bar{X} + 3s \\
&= 37 + (3 \times 4.92) \\
&= 51.8 \\
\text{LPL} &= \bar{X} - 3s \\
&= 37 - (3 \times 4.92) \\
&= 22.2
\end{aligned}$$

Note that

$$\begin{aligned}
\text{Process spread} &= \text{UPL} - \text{LPL} \\
&= 51.8 - 22.2 \\
&= 29.6
\end{aligned}$$

To project future performance, these calculations assume that the process is stable. We see in Problem 3.6 the process is unstable.

(b) The run chart appears below. It is not possible to evaluate process stability using a run chart.

3.6 The control chart values for the before period $\bar{X}$ chart are $\bar{\bar{X}} = 37.0$, UCL $= 41.4$, LCL $= 32.6$; for the R chart the values are $\bar{R} = 7.5$, UCL $= 15.9$. For the after period the values for the $\bar{X}$ chart are $\bar{\bar{X}} = 36.4$, UCL $= 41.0$, LCL $= 31.8$; R chart $\bar{R} = 7.9$, UCL $= 16.7$. The before period shows that the process is unstable and has numerous out-of-control signals. The after period shows that the process is reasonably stable, but more work is

3

needed to get below the 38% target. Note that the improvements have done little to reduce the mean or variability.

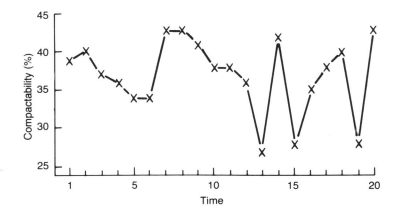

3.7 (a) The completed control charts appear as Charts 3.4 and 3.5. The process is out of control, as is indicated on both charts. Note that there are too many points in zone A and too few in zone C. The monitoring sampling plan is not too useful to help troubleshoot the causes for the instability in the process. Consecutive piece sampling should be used several times a shift for a few weeks. A monitoring plan can be used after the problem is resolved and stability verified.

(b) Computing $\hat{\sigma} = \bar{R}/d_2 = 1.9/2.06 = .9$. Process spread $= 6\hat{\sigma} = 5.5$. Note that since the process is not stable these values are not useful in predicting process performance.

(c) Since there are no points outside the specification limits, many people would consider the process acceptable. However, the lack of process stability means the future performance of the process is unpredictable. The state of process instability is unacceptable for a defect prevention system.

(d) There are 9 gage values between the specification limits but a minimum of 10 is preferred. It may be possible to enlarge the measurement scale to enable reading to the nearest half-thousandth so that values of 0, $\pm.5$, ±1.0, ±1.5, and so on, would be possible. This would give 17 scale divisions within the specification limits. Note that from Procedure 3.7 there are five possible values within the range control limits. This is a marginal situation so any reduction in variability may make the gaging inadequate.

3.8 (a) The control chart values for the $\bar{X}$ chart are $\bar{\bar{X}} = 28.3$, UCL $= 35.3$, LCL $= 21.3$; for the R chart the values are $\bar{R} = 12.1$, UCL $= 25.4$. Note that since a special cause was noted for point 34 it was omitted from the calculations. Several out-of-control signals can be noted on the $\bar{X}$ chart. However, the range chart appears stable.

(b) The process is not completely stable as evidenced by the out-of-control point, which required a tool change. Also, the process is quite variable (i.e., $\bar{R} = 12.1$) so a lack of process capability can be expected. Thus, sampling over a shorter time interval and collecting more subgroups would assist in identifying sources of variability.

(c) Both control limits are needed. Points below LCL indicate that the process has shifted to a significantly improved level. We would like to identify the special causes that are responsible for the improvement and implement them permanently!

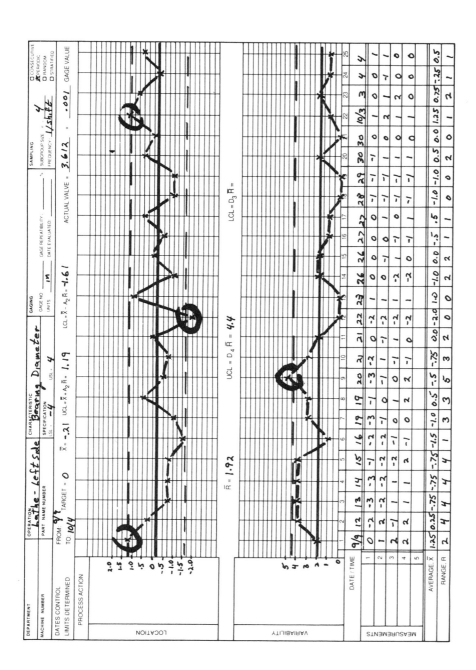

Chart 3.4 Control chart for variable data, Problem 3.7.

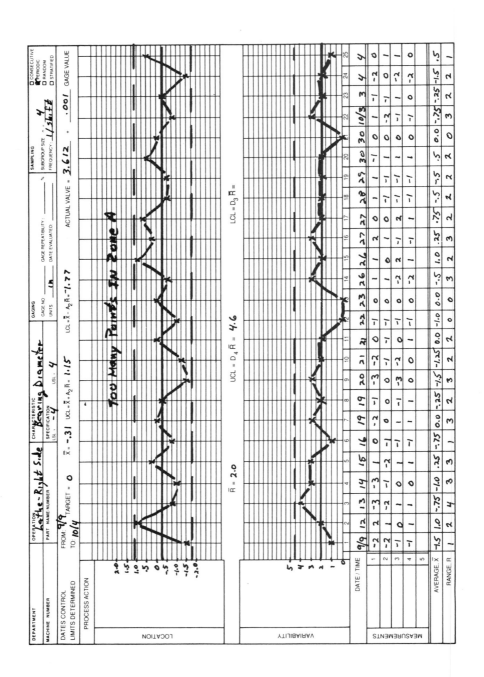

Chart 3.5 Control chart for variable data, Problem 3.7.

3.9 (a) The control chart values for the before period are $\bar{\bar{X}} = 14.54$, UCL $= 16.2$, LCL $= 12.9$; for the R chart the values are $\bar{R} = 2.85$, UCL $= 6.0$. The process is not stable since out-of-control signals appear on both the $\bar{X}$ and R charts as indicated in Chart 3.6. Also, there are too many values in zone A on both sides of $\bar{\bar{X}}$.

(b) The control chart values for the after period are $\bar{\bar{X}} = 15.2$, UCL $= 16.1$, LCL $= 14.3$; for the R chart the values are $\bar{R} = 1.6$, UCL $= 3.4$. Subgroup 5 was omitted from the calculation due to a known special cause. The process plotted on Chart 3.7 appears stable. However, there is a lack of measurement sensitivity that makes the control limits too narrow.

3.10 (a) The control chart values for the before period are $\bar{\bar{X}} = 46.6$, UCL $= 60.8$, LCL $= 32.5$; for the R chart the values are $\bar{R} = 24.4$, UCL $= 51.7$. The process is unstable for both location and variability. Several subgroups could have been excluded from the calculations, but this would make the out-of-control situation worse.

(b) The control chart values for the after period are $\bar{\bar{X}} = 21.4$, UCL $= 28.8$, LCL $= 14$; for the R chart the values are $\bar{R} = 12.8$, UCL $= 27.1$. The process still has an unstable mean, but the variability is now stable at a much reduced level. Further improvements are necessary, but progress is apparent.

3.11 (a) The table allows comparison of different sampling plans:

	Base $k = 25, n = 5$	Complete $k = 45, n = 5$	Complete $k = 75, n = 3$
$\bar{\bar{X}}$	11.6	11.4	11.4
$\mathrm{UCL}_{\bar{X}}$	14.8	14.7	16.1
$\mathrm{LCL}_{\bar{X}}$	8.4	8.1	6.7
$\bar{R}$	5.5	5.6	4.6
UCL_R	11.6	11.9	11.8
$\hat{\sigma}\ (\bar{R}/d_2)$	2.4	2.4	2.7

The process is probably not stable since there appear to be too few points in zones A and B, but with only $k = 25$ subgroups the pattern is not clear (see Chart 4.1).

(b) Using $k = 45$ subgroups does not change the control chart values appreciably. Normally, $k = 20$–25 subgroups provides stable control chart limits.

(c) Using a sampling plan of $n = 5$ when there are three major sources of variability is not advisable. Stratified control chart using $n = 3$ would be more appropriate for this situation (see Chap. 4). From the table, this chart results in an increased estimate of variability $\hat{\sigma}$. As Chart 4.2 indicates, this increase in variability results in a clearer indication of too few points in zones A and B.

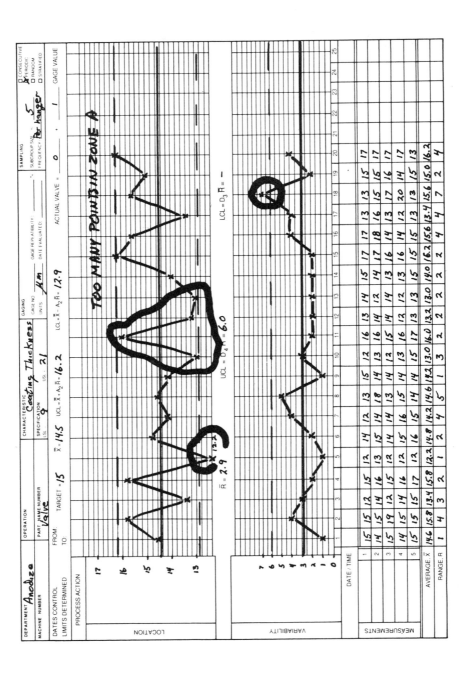

Chart 3.6 Control chart for variable data, Problem 3.9a.

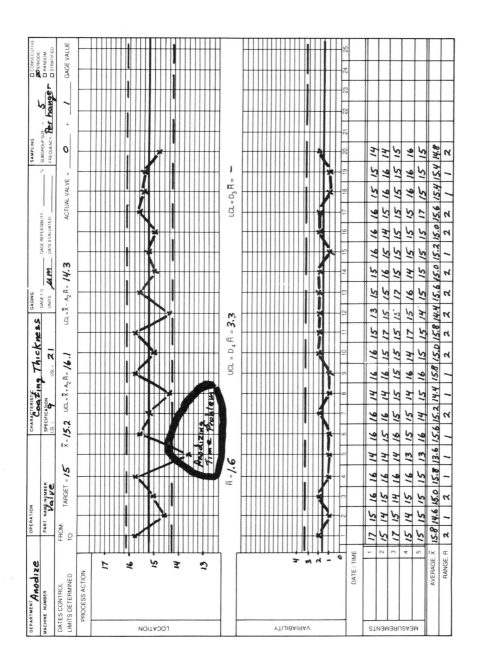

Chart 3.7 Control chart for variable data, Problem 3.9b.

3.12 (a) The control chart values for the 47 subgroups are, for the $\bar{X}$ chart, $\bar{\bar{X}} = 92.9$, UCL $= 113.1$, LCL $= 72.7$; for the R chart, $\bar{R} = 34.9$, UCL $= 74.0$. The process does not appear to be stable.

(b) The control chart values appear below:

			$\bar{X}$ Chart		R Chart	
		LCL	$\bar{\bar{X}}$	UCL	R	UCL
First	$k = 25$	70.7	89.3	107.9	32.0	67.8
Last	$k = 22$	74.7	96.9	119.1	38.2	81.0

The second period of $k = 22$ subgroups has a worse performance in both an increased delivery time and increased variability. The control chart appears in Figure 3.15.

3.13 (a) The control chart values for the position 1 median chart are $\bar{\bar{X}} = -2.0$, UCL $= -0.2$, LCL $= -3.9$; for the R chart the values are $\bar{R} = 2.7$, UCL $= 5.7$. For position 2 the values for the median chart are $\bar{\bar{X}} = -1.9$, UCL $= .0$, LCL $= -3.8$; R chart, $\bar{R} = 2.8$, UCL $= 5.9$. The process is clearly not stable, with subgroups 3, 6, 7, and 8 well above the other subgroups. However, the range appears reasonably stable. The process appears to be targeted low at about -2 to -4, perhaps to obtain smaller inner diameter (ID) bores to help address the loose fit problem.

(b) The measurement process consisted of placing the part on the gage and obtaining the measurement. It was possible to rotate the part on the gage and obtain a minimum ID and maximum ID as an indication of ovality. Controlling the bore ovality, which was defined as

Ovality $=$ maximum ID $-$ minimum ID

was found to be the solution to the loose fit problem. The wide range of variability within a single part is illustrated by preparing a histogram (Chap. 5) of the within-part variability.

Within-part variability = position 1 ID − position 2 ID

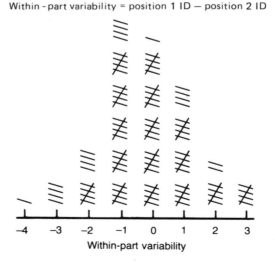

Within-part variability

Two points are worth noting: (1) an operational definition of ID size should be given (e.g., minimum ID, maximum ID, average ID), and (2) effective process control requires identification of the critical few characteristics that need to be controlled to produce the best possible products.

3.14 (a) The control chart values are computed for the hole using through the first subgroup on day 30 prior to the noted tool change:

	$\bar{X}$ Chart			R Chart	
	LCL	$\bar{\bar{X}}$	UCL	$\bar{R}$	UCL
Hole	6.7	7.3	7.9	.6	1.5
Tube	19.2	19.9	20.6	.7	1.7

The hole process is not stable, with several out-of-control $\bar{X}$ chart points and a long run above the mean at the end of the study. This behavior may be due to a number of unnoted tool changes (see shifting mean section in Chap. 4). The tube process is not stable, with several out-of-control points on the $\bar{X}$ chart and R chart.

(b) The mean for the hole is $\bar{\bar{X}} = 15.713$ mm, which differs from the 15.695 nominal by only .018 mm. The mean for the tube is $X = 15.839$ mm, which differs from the 15.825 nominal by only .014 mm. More importantly for fit, the difference in the means is .126 mm, which corresponds to a target difference of .13 mm. Thus, the two processes are, on the average, targeted well. However, the lack of stability would be expected to cause periodic assembly problems (tight fit) or potential leaks (loose fit).

3.15 The control chart values for a p chart are as follows:

$$\bar{p} = 9.4\%$$

$$\text{UCL} = \bar{p} + 3\sqrt{\frac{\bar{p}(100 - \bar{p})}{n}}$$

$$= 9.4 + 3\sqrt{8.52}$$

$$= 9.4 + 8.75$$

$$= 18.2\%$$

$$\text{LCL} = \bar{p} - 3\sqrt{\frac{\bar{p}(100 - \bar{p})}{n}}$$

$$= 9.4 - 8.75$$

$$= 0.7\%$$

The process is stable, as Chart 3.8 indicates.

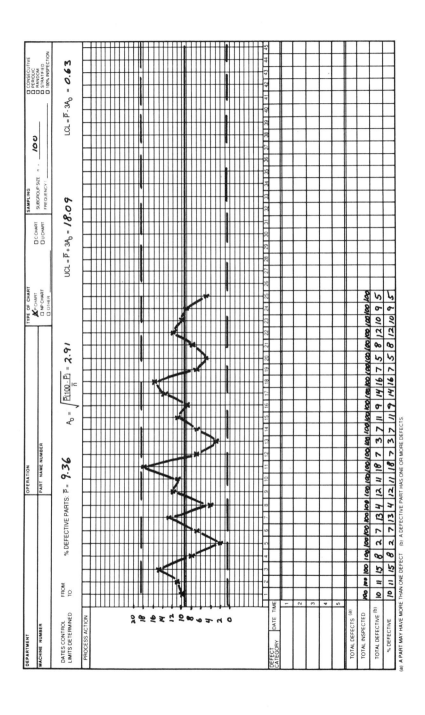

Chart 3.8 Control chart for attribute data, Problem 3.15.

3.16   (a) Values for the p charts are, for run 1,

$$\bar{p} = \frac{194}{375} = 51.7\%$$

UCL $= 90.4\%$

LCL $= 13.0\%$

and for run 2,

$$\bar{p} = \frac{196}{375} = 52.3\%$$

UCL $= 91.0\%$

LCL $= 13.6\%$

The $k = 25$ subgroups produce results that are very consistent. The coin tossing process is stable.

(b) Highest value $= 15 \times .91 = 13.6$ or 13; lowest value $= 15 \times .136 = 2.0$ or 2.

3.17   (a) The control chart values are, for the before period, $\bar{p} = 138/12{,}500 = 1.1\%$, UCL $= 2.5\%$, LCL $= 0\%$. The process is stable except for 1 day, where no special cause was noted. The control chart values are, for the after period, $\bar{p} = .11\%$, UCL $= .55\%$, LCL $= 0\%$. These calculations omit the day on which a special cause was noted.

(b) The team concluded that since the process was reasonably stable the only way to reduce the overall level of defects was to change the process (i.e., the system).

3.18   The control chart values are $\bar{p} = 0.21$, UCL $= 0.47$, and LCL $= 0$. Since the asterisk denoted a different extension housing model, these values were omitted from the calculations. The average sample size was $\bar{n} = 2781$, which has a 25% interval of (2086, 3476). The control limits were calculated using $\bar{n}$; points with n_i outside the 25% interval were calculated using the individual n_i values. The resulting graph for the last 25 points appears in Chart 3.9. The process is not completely stable, with several days indicating the presence of special causes. Clearly, the different model results in a much higher reject rate.

3.19   (a) The subgroups 1–8 control limit should be calculated separately from 9 to 25 since a different rejection criterion is being used. Subgroups 21 and 22 should be omitted because of an identified special cause. The control chart values will not be good estimates since fewer than 20 subgroups will be used for both sets of subgroups.

(b) The control chart values are shown on Chart 3.10. Note that different control limits are used for each subgroup since the subgroup sizes vary. The process is clearly unstable.

3.20   Control chart values are

$$\bar{c} = 5.2$$

$$\text{UCL}_c = 5.2 + 3\sqrt{5.2} = 12$$

$$\text{LCL}_c = 0$$

The process is stable.

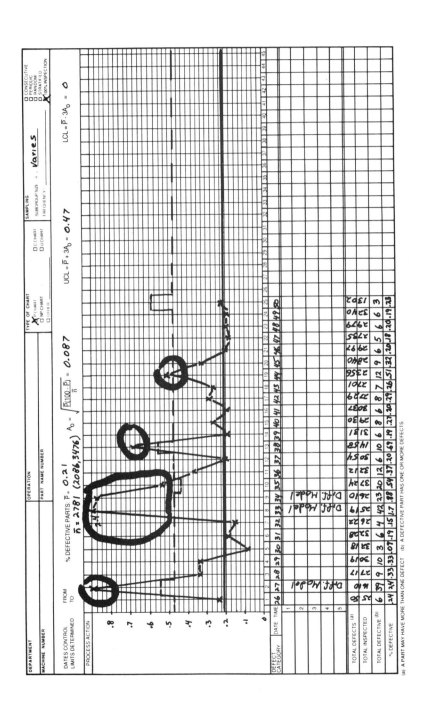

Chart 3.9 Control chart for attribute data, Problem 3.18.

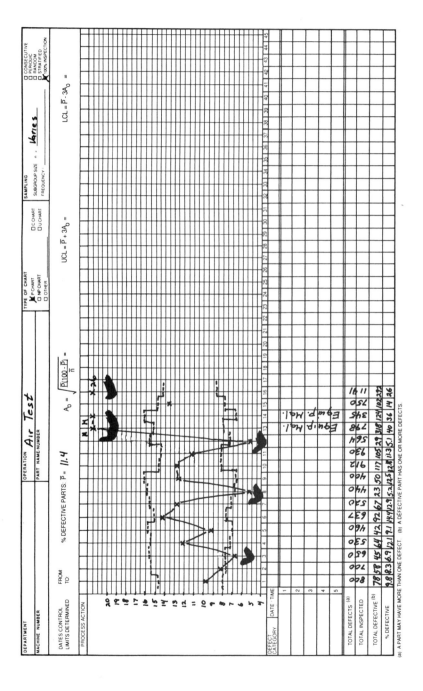

Chart 3.10   Control chart for attribute data, Problem 3.19.

Solutions to the Problems in Chapter 4

4.1 (a) The control chart values for the $\bar{X}$ chart are $\bar{\bar{X}} = 1.96$, UCL $= 3.07$, LCL $= 0.85$; for the R chart, $\bar{R} = 3.0$, UCL $= 5.6$, LCL $= 0.4$. One might conclude from Figure 4.15 that the control chart appears generaliy stable except for a single out-of-control subgroup on the $\bar{X}$ chart. This conclusion is, however, incorrect as part b and Problem 5.8 show.

(b) The means and standard deviations for the spindles are as follows:

	Spindle							
	1	2	3	4	5	6	7	8
$\bar{X}$	1.9	2.5	2.2	2.4	1.4	1.9	1.3	2.0
s	1.1	1.2	1.1	1.3	0.9	1.1	0.7	1.4

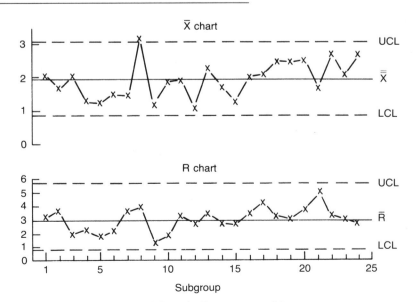

Figure 4.15 Control chart for eight-spindle screw machine.

17

The means range between 2.5 and 1.3, and the standard deviations range between 1.4 and 0.7. These differences appear somewhat larger than might be expected (Prob. 11.5 will give a more detailed analysis method).

(c) This is a reasonable application of a stratified control chart; however, two different charts using $n = 4$ may be easier to manage. It appears from b that a baseline study was not performed to ensure that the spindle means and standard deviations were about the same.

4.2 (a) The control chart values for the $\bar{X}$ chart are $\bar{\bar{X}} = 63.9$, UCL = 76.1, and LCL = 51.7; for the R chart, $\bar{R} = 21.0$, UCL = 44.5. The $\bar{X}$ and R charts both show a lack of stability since a pattern exists in which both measurements for a shift have similar values. This pattern could be due to operators from different shifts not using the same procedure to make measurements. Generally, shift 1 has lower values than shift 2. This pattern could also result from a tool change after the first shift. We cannot be sure of the special cause without more information.

(b) The current sampling plan was successful in identifying a lack of stability. However, because of the long time interval covered by a subgroup, it will probably not be successful in troubleshooting to identify special causes. A good plan would be to temporarily use consecutive piece sampling (say, one subgroup per hour) with the objective of finding and eliminating any special causes.

4.3 (a and b) The control chart values are as follows:

Position	$\bar{X}$ Chart			R Chart	
	LCL	$\bar{\bar{X}}$	UCL	$\bar{R}$	UCL
A	29.7	36.6	43.5	6.8	17.5
B	29.8	36.6	43.4	6.7	17.3
C	30.1	35.4	40.7	5.2	13.4

The process at positions A and B appears to be stable, but a shift in the mean level occurs around subgroup 15 at position C.

(c) Note that the means at positions B and C appear to differ and position C has lower variability $\bar{R}$ and thus tighter control. The logical question to ask is how we can reduce the variability at B to be at least as low as at C. The means of both processes should be centered at 42 to minimize an out-of-specification condition.

(d) The sampling plan at B and C would, in general, be more likely to show an out-of-control condition since the fewer sources of variability result in tighter control limits. In the troubleshooting mode of operation, out-of-control signals are desirable since they, when pursued, may help identify special causes of variation. Elimination of these problems will reduce overall process variability and make process stability more likely.

4.4 (a) The adjustments to the process are shown below.

Stable process	13	11	10	8	11	10	9	11	9	8
Adjusted process	—	8	9	7	13	9	8	12	8	9
Adjustment	−3	+2	—	+3	−3	—	+2	−2	+2	—
Accumulated adjustment	−3	−1	−1	+2	−1	−1	+1	−1	+1	+1

Stable process	11	8	10	10	11	11	10	7	10	10
Adjusted process	12	7	12	10	11	11	10	7	13	10
Adjustment	−2	+3	−2	—	—	—	—	+3	−3	—
Accumulated adjustment	−1	+2	0	0	0	0	0	+3	0	0

Stable process	9	7	10	11	10	7	10	8	9	10
Adjusted process	9	7	13	11	10	7	13	8	11	12
Adjustment	—	+3	−3	—	—	+3	−3	+2	—	−2
Accumulated adjustment	0	+3	0	0	0	+3	0	+2	+2	0

The means and standard deviations are as follows:

	$\bar{X}$	s
Stable process	9.63	1.43
Adjusted process	9.90	2.08

The adjusted process has 45% more variability than the stable process.

(b) The process requires 18 adjustments, which is greater than 8 because we compensate for a high or low value and must recompensate to correct for the unneeded adjustment.

4.5 (a) The run chart appears in Figure 4.16. The wild swings in the process suggest that the compensation system is attempting to adjust for special causes of variation. This

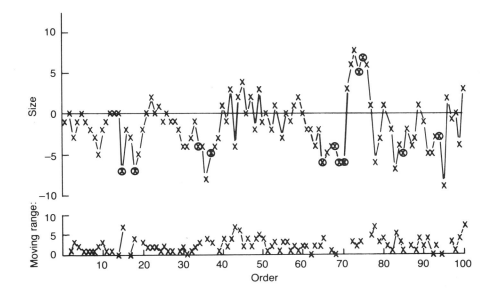

Figure 4.16 Run chart for automatic compensation system.

Table 4.4   Moving Range Calculations for Problem 4.5

No.	Value	Range	No.	Value	Range	No.	Value	Range	No.	Value	Range
1	−1		26	0	1	51	0	1	76	6	
2	0	1	27	−1	1	52	−2	2	77	1	5
3	−3	3	28	−1	0	53	1	3	78	−6	7
4	−1	2	29	−2	1	54	0	1	79	−3	3
5	0	1	30	−4	2	55	−3	3	80	1	4
6	−1	1	31	−4	0	56	0	3	81	−1	2
7	−2	1	32	−3	1	57	−1	1	82	−2	1
8	−3	1	33	−1	2	58	1	2	83	−7	5
9	−5	2	34	−4A	3	59	2	1	·84	−4	3
10	−2	3	35	−4		60	0	2	85	−5A	1
11	−1	1	36	−8	4	61	−2	2	86	−2	
12	0	1	37	−5A	3	62	−2	0	87	−4	2
13	0	0	38	−4		63	−4	2	88	−3	1
14	0	0	39	−3	1	64	−2	2	89	1	4
15	−7A	7	40	1	4	65	−6A	4	90	−1	2
16	−2		41	−1	2	66	−5		91	−5	4
17	−3	1	42	3	4	67	−4	1	92	−5	0
18	−7A	4	43	−4	7	68	−4A	0	93	−3	2
19	−5		44	2	6	69	−6A		94	−3A	0
20	−2	3	45	4	2	70	−6A		95	−9A	
21	0	2	46	0	4	71	3		96	2	
22	2	2	47	2	2	72	6	3	97	−1	3
23	0	2	48	−2	4	73	8	2	98	0	1
24	1	1	49	3	5	74	5S	3	99	−4	4
25	−1	2	50	−1	4	75	7S		100	3	7

practice will add to the confusion of trying to troubleshoot the system. Note that the variability may be due to the machining system or the gaging system required to perform the compensation.

(b) A moving range can be computed by obtaining the range of two successive values when no compensation occurred. Table 4.4 contains the moving range calculations. The R chart values are $\bar{R} = 2.36$ for 86 values with UCL = 7.7, which uses a subgroup size $n = 2$ (i.e., range computed from two successive values). The apparent stable situation is worse than indicated on the R chart in Figure 4.16 since a number of points associated with a large change were part of a compensation so no range could be calculated (e.g., point 95).

4.6   (a) Figure 4.17 shows the control chart results were $\bar{\bar{X}} = −.04$, UCL = 1.4, and LCL = $−1.5$ for the $\bar{X}$ chart and $\bar{R} = 2.5$ and UCL = 5.3 for the R chart. It is apparent that this is a standard tool wear application with an increasing mean. The usual $\bar{X}$ chart is not meaningful for controlling this process. Note, however, that the R chart is stable and provides a good method of controlling process variability.

(b) A reasonable approach to controlling the changing mean is to use a modified control limit, the R chart provides a stable estimate of

$$\hat{\sigma} = \frac{\bar{R}}{d_2} = \frac{2.5}{2.33} = 1.07$$

The rejection limits are

$$\text{URL} = 10 - 0.91(2.5) = 7.7$$
$$\text{LRL} = -10 + 0.91(2.5) = -7.7$$

From the technical viewpoint these limits could be used with little danger of producing parts beyond the specification limits. However, two other factors should be evaluated. First, only one tool was studied. It is possible that other tools will have a different wear characteristic so a conservative approach is warranted. Second, the time between tool changes is quite long. In this study the same tool was used for all 25 subgroups (collected about 1 hour apart), so 25–30 hours of operation are possible to be within URL and LRL. Although this approach minimizes tool costs, a significant opportunity is lost. By changing the tool once an 8 hour shift, a change of less than about 5 units could be expected. Thus, a system in which a tool is routinely changed once a shift is not only easier to manage but uses only about 25% of the specification range. Clearly, these parts would be more likely to meet customers' needs than parts produced by a system with a 3 day tool change cycle.

4.7   (a) The control chart values for the $\bar{X}$ chart are $\bar{\bar{X}} = 1.6$, UCL $= 2.5$, and LCL $= 0.7$; for the R chart, $\bar{R} = 1.5$, UCL $= 3.3$. The R chart is stable, but the $\bar{X}$ chart shows several out-of-control points.

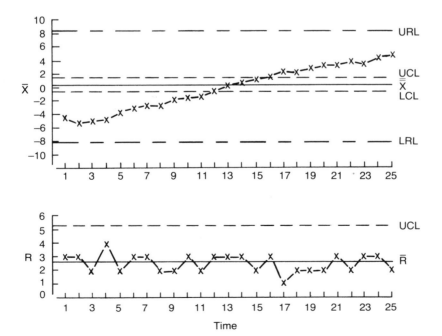

Figure 4.17   Tool wear control chart.

(b) Since the range is stable, the process should be able to be targeted at 1.0, which suggests that the process is being run too high. The process is easily adjustable by the operators. The pattern of out-of-control signals suggests that there are too many adjustments taking place, with different operators selecting their own target values.

(c) There are only 4 units of measurement within the specification range. Thus, by almost any criterion the measurement system is considered to lack adequate sensitivity. However, two factors may make additional gaging unnecessary. First, the function of the outer diameter (OD) size of the part needs to be considered. Would significantly improving the gaging and capability of the process improve the function or durability of the gear/pin interface? If the answer is yes, new gaging may be warranted. Second, the units of measurement are .0001 inch, and improving the precision may not only be costly, but more sophisticated gages may not be able to withstand the manufacturing floor environment or may be difficult for operators to use.

4.8 (a) Chart 4.7 shows the control chart results were $\bar{\bar{X}} = 3.94$, UCL $= 3.99$, and LCL $= 3.90$ for the $\bar{X}$ chart and $\bar{R} = 0.15$, UCL $= 0.27$, and LCL $= .03$ for the R chart. It is apparent that there are too many points in zone A and that there is an out-of-control point on both the $\bar{X}$ and R charts. These out-of-control signals may be due to inadequate measurement sensitivity since there are only three possible values within the range control limits.

(b) One method of addressing low measurement variability is to increase the subgroup size, which has not proven completely effective in this case. A preferable approach is to spread out the sampling so that all samples in a subgroup are not collected over a short period of time. In a batch-type process there often will be minimal variability over a short period of time (i.e., within a batch). Perhaps collecting samples every 15 minutes would represent the process better. The drawback of this approach is that the time required for collecting samples increases dramatically. Also, if a subgroup size of $n = 4$ is used with this approach, 1 hour would need to pass before an out-of-control condition could be discovered. Ideally, upstream controls should be used to control the process and detect any lack of stability.

4.9 The $\bar{X}$ chart values are $\bar{\bar{X}} = 7.8$, UCL $= 36.1$, LCL $= -20.5$; the R chart values are $\bar{R} = 38.8$, UCL $= 88.5$. The process is stable, with the exception of point 24, which has an out-of-control range. The major problem appears to be the high range of difference between the two procedures. This result indicates that the production method is much too variable. Also, we would want the mean 7.8% difference to be closer to 0% so that there would be no bias in the quick method. These results prompted a team to address the setup procedure and correct these problems.

4.10 (a) The control chart is given in Chart 4.8. The process appears stable.

(b) There are two reasons. First, by spreading the sample out uniformly for each division there is reasonable assurance that the subgroup represents all parts in the division adequately. Second, spreading the samples out enables more variability to be observed, which minimizes the influence of the lack of measurement sensitivity (i.e., $\bar{R}$ is 1.4, only about 1 unit).

(c) Although there is a lack of measurement sensitivity, the process is very capable since the specification is 20–40. Note, however, that the process is not centered. Because there are only four possible range values within the range control limits, some out-of-control signals can be expected due to only the lack of measurement sensitivity. A more sensitive measurement method would make a better chart, but the current method seems adequate.

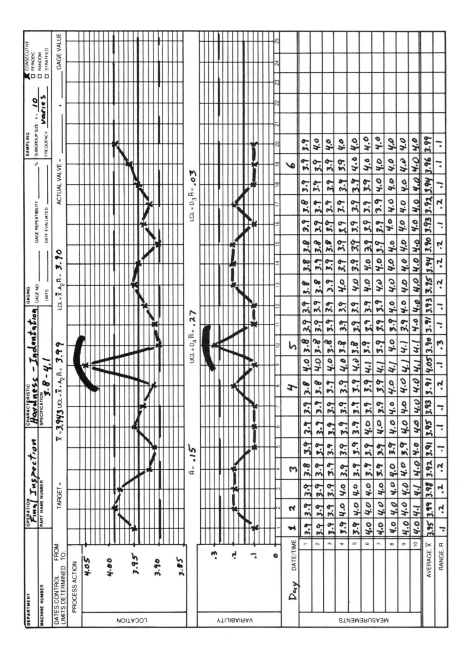

Chart 4.7   Control chart for variable data, Problem 4.8.

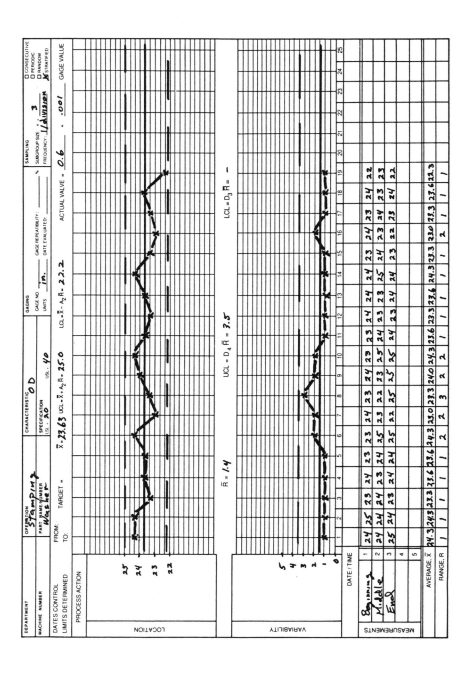

Chart 4.8 Control chart for variable data, Problem 4.10.

4.11 (a) The difference and the subgroup means and ranges are listed below:

Part	Differences			$\bar{X}$	R
1	−.0016	−.0016	.0020	−.00040	.0036
2	−.0009	.0010	−.0002	−.00003	.0019
3	−.0010	−.0028	−.0004	−.00140	.0024
4	.0021	.0020	−.0014	.00090	.0035
5	−.0013	−.0013	.0009	−.00057	.0022
6	−.0022	.0011	−.0007	−.00060	.0033
7	−.0024	−.0008	−.0014	−.00153	.0016
8	−.0009	−.0021	.0043	.00043	.0064
9	−.0023	−.0009	.0000	−.00107	.0023
10	−.0008	−.0032	.0002	−.00127	.0034
11	−.0009	−.0004	.0005	−.00027	.0014
12	−.0000	−.0014	−.0019	−.00110	.0019
13	−.0010	−.0015	.0013	−.00040	.0028
14	−.0004	−.0003	.0005	−.00007	.0009
15	.0001	.0006	.0010	.00057	.0009
16	−.0013	.0000	.0020	.00023	.0033
17	.0000	.0010	−.0008	.00007	.0018
18	−.0015	−.0013	.0000	−.00093	.0015
19	−.0012	−.0001	−.0004	−.00057	.0011
20	.0006	−.0002	−.0007	−.00010	.0013
		$\bar{\bar{X}} = -.00041$		$\bar{R} = .00237$	

$$\text{UCL}_R = D_4\bar{R} = 2.58 \times .00237 = .006$$
$$\text{UCL}_{\bar{X}} = \bar{\bar{X}} + A_2\bar{R} = -.00041 + (1.02 \times .00237) = .0020$$
$$\text{LCL}_{\bar{X}} = \bar{\bar{X}} - A_2\bar{R} = -.00041 - (1.02 \times .00237) = -.0028$$

The process is not completely stable since part 8 has an out-of-control range.
  (b) Note that $\hat{\sigma} = \bar{R}/d_2 = .00237/1.69 = .0014$, so the process spread is

$$\text{Process spread} = \frac{6\bar{R}}{d_2} = .0084 \text{ inch}$$

so the spread of the process is greater than the stated spread of the machine (.002 inch).

Solutions to the Problems in Chapter 5

5.1 The histogram appears below:

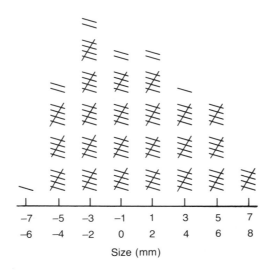

Several interpretations are possible. The pattern of measurements is smooth but clearly not bell shaped. This could be a natural characteristic of the process; there is no "rule" that a process must have a bell-shaped distribution. However, many size characteristics do exhibit a bell-shaped distribution. The clifflike feature on the left side of the histogram does look suspicious for a size characteristic and could be due to a measurement or recording problem. Process knowledge must be used to clarify the interpretation.

5.2 (a) The stem and leaf plot appears below:

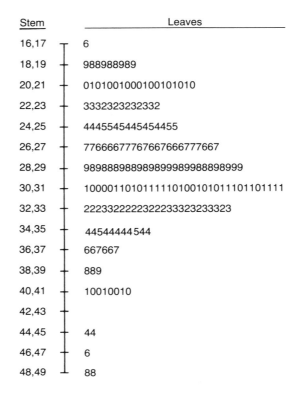

Stem	Leaves
16,17	6
18,19	988988989
20,21	01010010001010101010
22,23	3332323232332
24,25	4445545445454455
26,27	77666677767667666777667
28,29	98988898898989998988898999
30,31	1000011010111110100101011101101111
32,33	2223322222322233323233323
34,35	44544444544
36,37	667667
38,39	889
40,41	10010010
42,43	
44,45	44
46,47	6
48,49	88

(b) The plot makes preparation of the histogram of the original data easy since the number of measurements in an interval can be easily counted. This plot is also useful in preparing the histogram of the log of the measurements. Using the scale of original and logs in Example 5.5, the number of values in each interval can be determined. For example, the number of log measurements in the 3.55–3.65 interval, which include original measurements 35, 36, 37, and 38, is the sum of 2 (35), 4 (36), 2 (37), and 2 (38), or a total of 10.

(c) The histogram also indicates the special-cause problem requiring a tool change by the outlier values. The general shape is typical of surface finish distributions.

5.3 A histogram for each side appears below.

Left side:

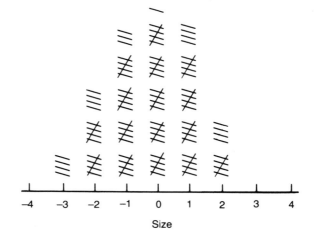

Right side:

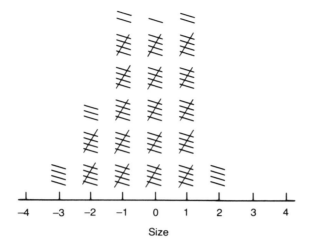

The two sides appear to have about the same size location and variability. Both histograms are reasonably smooth. It does seem somewhat unusual that there are no measurements in the extremes near the ±4 specification limits. It is not clear whether measurements are not being accurately reported; further investigation is necessary.

5.4 The histogram for charts 4.4, 4.5, and 4.6 appear below.

From Chart 4.4:

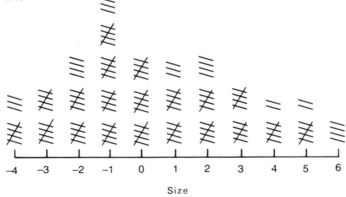

From Chart 4.5, spindles 8, 6, 4, and 2:

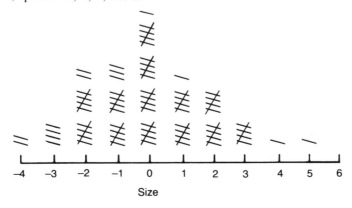

From Chart 4.5, spindles 1, 7, 5, and 3:

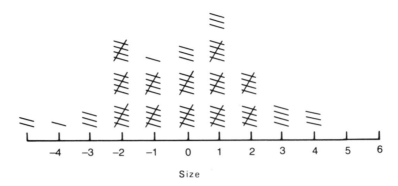

The top histogram appears to have a clifflike shape on the left. Perhaps data below −4 were reported as −4 since the target was −4 to 3. The upper values also appear to have a clifflike shape. The second histogram appears to be a mixture, possibly due to spindles 3 and 5 being off target. The bottom histogram does not appear unusual even though Chart 4.6 indicated two spindles with off-target means. Spindle 8 being high and spindle 2 low may have an offsetting influence on the histogram's shape. An individuals histogram for the spindles should be plotted.

5.5   (a) The histograms are shown below.

Before study:

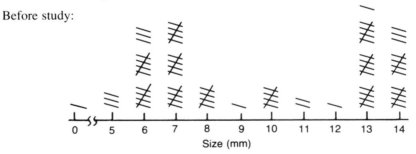

After study:

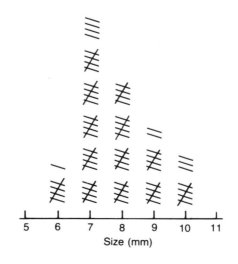

It is apparent that the before study had a significant process problem and that the coolant addition greatly improved the process spread. However, the clifflike shape of the before histogram is curious. Since the "cliff" occurs at the specification limit, there appears to be sorting or misrecording taking place.

(b) Calculating the means and standard deviation using the steps in Procedure 5.2 gives the following result. Step 1: The frequency calculations are as follows:

Interval	x	f	xf	fxx
1	6	6	36	216
2	7	29	203	1421
3	8	20	160	1280
4	9	12	108	972
5	10	8	80	800
		$N = 75$	$A = 587$	$B = 4689$

Step 2:

$$\bar{X} = \frac{587}{75} = 7.82$$

Step 3:

$$E = \frac{(587)(587)}{75} = 4594.25$$

Step 4:

$$\frac{B - E}{N - 1} = \frac{4689 - 4594.25}{74} = 1.280$$

Step 5:

$$s = \sqrt{1.280} = 1.13$$

The upper and lower process limits are as follows:

UPL $= \bar{X} + 3s = 7.82 + (3 \times 1.13) = 11.2$
LPL $= \bar{X} - 3s = 7.82 - (3 \times 1.13) = 4.4$

To compute the process limits in the actual dimension scale, we substitute into the equation

Actual value $= 1.1984 + .0001 \times$ coded value
Actual UPL $= 1.1984 + (.0001 \times 11.2)$
$\qquad = 1.19952$
Actual LPL $= 1.1984 + (.0001 \times 4.4)$
$\qquad = 1.19884$

Note that it is not meaningful to perform these calculations on the before measurements. The unusual shape of the histogram and likely unstable process makes $\bar{X}$ and s not meaningful.

5.6 (a) The histograms for the before and after measurements are as follows.

Before:

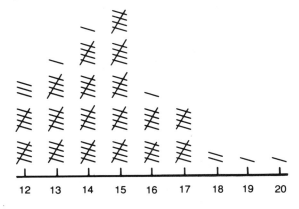

After:

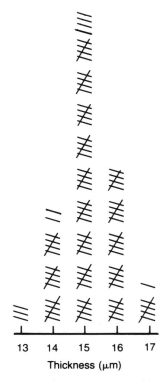

The improvement in the after period is apparent. It is interesting that the before period exhibits a clifflike histogram. The cause for this result was not known.

(b) The histograms for the before and after measurements are as follows.

Before:

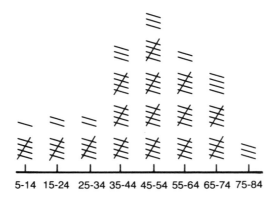

5-14 15-24   25-34 35-44 45-54 55-64 65-74 75-84

After:

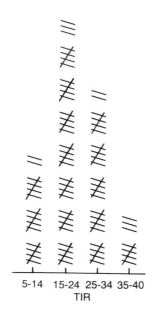

5-14   15-24 25-34 35-40
TIR

Again the improvement is apparent. The after process was still not stable (Chap. 3 analysis), but the reduced variability and lower TIR is an improvement.

5.7 (a) The histograms of individual machines are as follows:

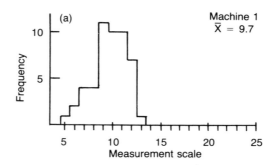

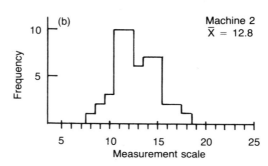

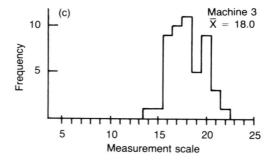

The increasing means from machines 1–3 are apparent. (b) The histograms of all combined machines are as follows:

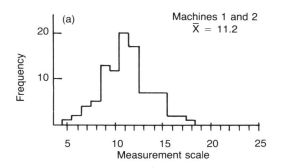

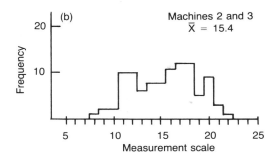

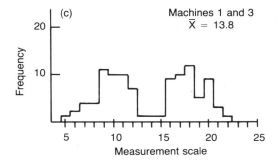

There is little evidence of two populations in the combined machine 1 and 2 histogram. (c) The combined histogram for machines 2 and 3 suggests two possible populations. (d) The combined histogram for machines 1 and 3 clearly shows two populations. (e) The combined histogram for all machines indicates two (not three) populations:

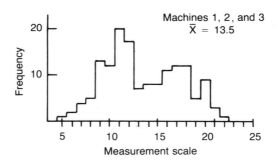

5.8   The histogram of the screw machine data appears below:

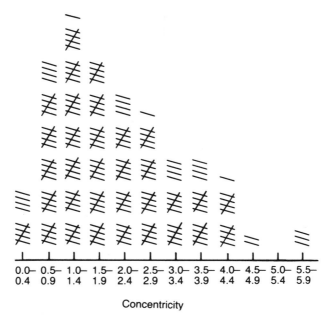

Concentricity

The histogram has a heavy right tail, which is the result of having varying spindle means, as was indicated in Problem 4.1b.

5.9   (a) The histogram for the $N = 125$ values appears below:

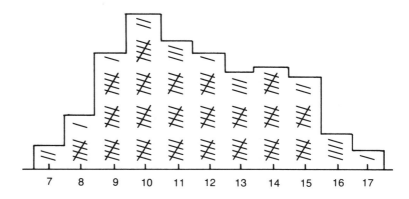

This histogram is clearly not bell shaped, with too many high values. Recall that this was due to spindle 2 having a higher mean than spindles 1 and 3.

(b) The $N = 125$ histogram appears very similar to the $N = 225$ and is adequate for interpretation. The $N = 225$ histogram gives a clearer picture of the process, but the extra effort to evaluate 100 more parts would likely not be worthwhile.

(c) The zones for the individual histograms are as follows:

$$\bar{\bar{X}} + 3\hat{\sigma} = 11.6 + (3 \times 2.4) = 18.8$$
$$\bar{\bar{X}} + 2\hat{\sigma} = 11.6 + (2 \times 2.4) = 16.4 \quad \leftarrow \text{Zone A}$$
$$\bar{\bar{X}} + 1\hat{\sigma} = 11.6 + (1 \times 2.4) = 14.0 \quad \leftarrow \text{Zone B}$$
$$\bar{\bar{X}} = 11.6 \quad \leftarrow \text{Zone C}$$
$$\bar{\bar{X}} - 1\hat{\sigma} = 11.6 - (1 \times 2.4) = 9.2 \quad \leftarrow \text{Zone C}$$
$$\bar{\bar{X}} - 2\hat{\sigma} = 11.6 - (2 \times 2.4) = 6.8 \quad \leftarrow \text{Zone B}$$
$$\bar{\bar{X}} - 3\hat{\sigma} = 11.6 - (3 \times 2.4) = 4.4 \quad \leftarrow \text{Zone A}$$

(d) The calculation table appears below:

1.

Interval	x	f	xf	xx	fxx
1	7	2	14	49	98
2	8	6	48	64	384
3	9	16	144	81	1296
4	10	22	220	100	2200
5	11	18	198	121	2178
6	12	16	192	144	2304
7	13	13	169	169	2197
8	14	15	210	196	2940
9	15	12	180	225	2700
10	16	4	64	256	1024
11	17	1	17	289	289
Total		$N = 125$	$A = 1456$		$B = 17{,}610$

2. $\quad \bar{X} = \dfrac{A}{N} = \dfrac{1456}{125} = 11.65$

3. $\quad E = \dfrac{(1456)(1456)}{125} = 16{,}959.49$

4. $\quad \dfrac{B - E}{N - 1} = \dfrac{17{,}610 - 16{,}959.49}{124} = 5.246$

5. $\quad s = \sqrt{5.246} = 2.29$

Typically, s is a better estimate of the standard deviation than $\hat{\sigma}$ computed from control charts. However, in this case the two values agree very well. In this situation, we would not expect close agreement since the histogram is not normal and bell shaped.

(e) The following computations use the expected percentage in each zone to compute the expected number of samples in the zone (expected $\%N$ = expected no.). These values can then be compared to the actual numbers in the zones.

Zones	Upper interval	Expected %	Expected no. $N = 125$	Actual no.
A high	18.8	2	2.5	1
B high	16.4	14	17.5	16
C high	14.0	34	42.5	44
C low	9.2	34	42.5	40
B low	6.8	14	17.5	24
A low	4.4	2	2.5	0

Notice that there is not a clear indication of a mixture problem using the zones test for the histogram than with a control chart. This is similar to the control chart analysis.

5.10   (a) The control charts appear as Charts 5.2 and 5.3. There is an inverse relationship between hardness and indentation diameter, so the out-of-control signals are a mirror image of each other. The two measurement scales yield similar results.

(b) A histogram using both scales of measurement appears below. Unfortunately, it is not possible to determine the best scale for two reasons. First, the process is not stable so a bell-shaped histogram would not be expected. Second, the number of actual measurement units is 6 (or 4 for the main group of data). Thus, there is insufficient spread in the histogram to evaluate the best approach. The lack of measurement sensitivity is also apparent in the range chart with only 4 possible values within the control limits (Chart 5.3). Given no preference of scales, the original measurement scale (diameter) is probably more useful in this case.

(c) This sampling problem is typical of batch industries that have many short production runs. The long span of time between some subgroups suggests that the chart does not really effectively evaluate process stability. Upon investigation, it was determined that there were really only three target ranges for hardness run in the foundry. Thus, the foundry process was to control three processes, but metal would be poured in molds creating 100 or more parts. A control chart for each part was not only quite difficult to maintain, but largely meaningless. A better approach would be to maintain three charts, one for each hardness target value. Many parts might appear on a single chart. The subgroup size and frequency of sampling would need to be adapted to the process. Notice that even this sampling process that was used showed that the part met the specification but showed a lack of control. Troubleshooting will require a more meaningful sampling plan.

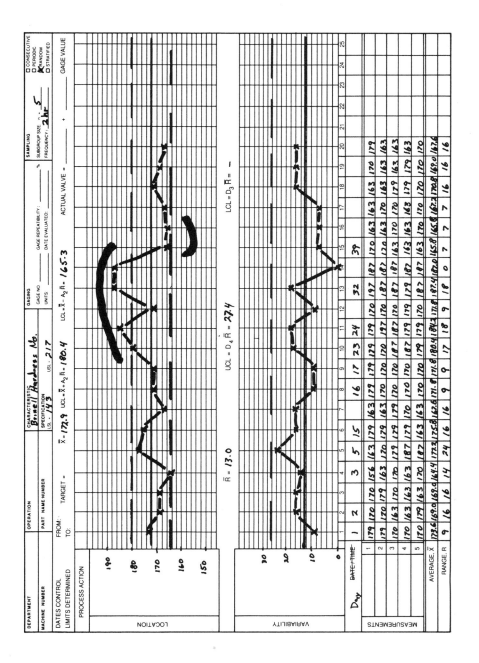

Chart 5.2 Control chart for variable data, Problem 5.10.

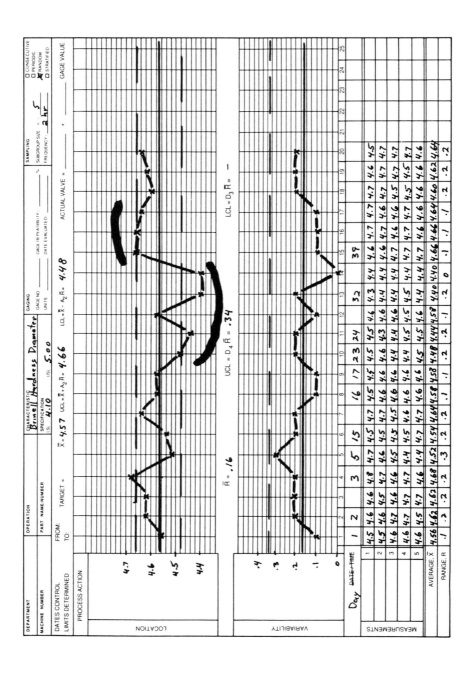

Chart 5.3   Control chart for variable data, Problem 5.10.

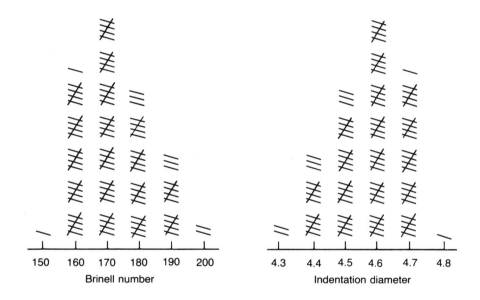

Brinell number   Indentation diameter

Solutions to the Problems in Chapter 6

6.1 The R chart shows a run of eight points below the mean for subgroups 9–16, suggesting a decrease in process variability. However, the out-of-control signal for subgroups 23–25 indicates increased process variability. The $\bar{X}$ chart shows a run of 11 points above the mean for subgroups 1–11, indicating increased location. Subgroups 12–16 indicate a decrease in location. Subgroups 18–25 again indicate an increase in mean level.

The actual simulation used the following population:

Subgroup	Population	Mean	SD (σ)
1–11	B	2	1.74
12–16	E	−4	1.74
17–25	H	2	2.35

The control chart indicated the correct process changes, with the exception of the increased variability for subgroups 17–22.

6.2 The range chart indicates several subgroups beyond the control limits and a significant number of high range values. The $\bar{X}$ chart shows a long run of values above the mean for subgroups 28–37 and a significant increase for subgroups 44–45. Visually scanning the rows indicates an increase mean for spindle 2 for subgroups 26–37 and increased variability for spindle 3 for subgroups 38–50. The quick test confirms this pattern.

The actual simulation used the following populations:

Subgroups	Spindle	Population	Mean	SD (σ)
26–37	1	A	0	1.74
	2	C	4	1.74
	3	A	0	1.74
38–50	1	A	0	1.74
	2	A	0	1.74
	3	F	0	3.74

The control chart correctly indicated the process changes.

Solutions to the Problems in Chapter 7

7.1

Population	$\bar{X}$	σ	C_p	CPU	CPL	k	C_{pk}
A	0	1.74	1.92	1.92	1.92	0	1.92
B	2	1.74	1.92	1.53	2.30	0.2	1.53
C	4	1.74	1.92	1.15	2.68	0.4	1.15
D	−2	1.74	1.92	2.30	1.53	0.2	1.53
E	−4	1.74	1.92	2.68	1.15	0.4	1.15
F	0	3.74	0.89	0.89	0.89	0	0.89
G	0	2.35	1.42	1.42	1.42	0	1.42
H	2	2.35	1.42	1.13	1.70	0.2	1.13

7.2 Current capability is

$$C_p = \frac{18 - 10}{6 \times 1} = 1.33 \qquad C_{pk} = \text{CPU} = \frac{18 - 16}{3 \times 1} = 0.67$$

The spread $\bar{X} + 3\hat{\sigma}$ must equal USL, so

$$\bar{X} + 3\hat{\sigma} = \text{USL}$$
$$\bar{X} = \text{USL} - 3\hat{\sigma}$$
$$= 18 - (3 \times 1)$$
$$= 15$$

To test the result,

$$C_{pk} = \text{CPU} = \frac{18 - 15}{3 \times 1} = 1.0$$

Setting $\bar{X} = 16$ and USL $= 18$ in this equation gives $\hat{\sigma} = .67$.

7.3 A histogram of the data is given below:

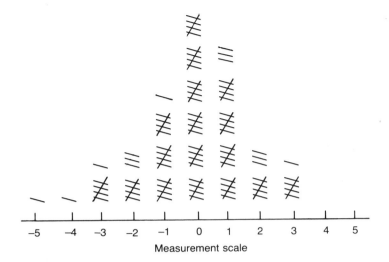

Measurement scale

The process capability work sheet is completed on the next pages. The standard deviation is 1.63, which is close to the $\hat{\sigma}$ value of 1.73. This is expected when a process from a normal distribution is in statistical control.

Process Capability Work Sheet

Process specifications:

Upper specification limit:	USL = 6
Lower specification limit:	LSL = −6
Width of specification:	$W = \text{USL} - \text{LSL} = 6 - (-6) = 12$
Midpoint of specification:	$M = \dfrac{\text{USL} + \text{LSL}}{2} = \dfrac{6 + (-6)}{2} = 0$

Process variability (two methods):

Control chart	$\bar{R} = 3.56$
Standard deviation of individuals	$\hat{\sigma} = \dfrac{\bar{R}}{d_2} = \dfrac{3.56}{2.06} = 1.73$
Histogram	$s = 1.61$

Note that histograms do not establish statistical control or predictability of a process. Histograms and s should only be used to identify a current problem process.

Process spread   PS = $6\hat{\sigma}$ =

PS = $6s = 6 \times 1.61 = 9.66$

Note that $\hat{\sigma}$ and s should be approximately equal for a normal process that is in statistical control.

Process location:

Process mean $\quad\quad\quad \bar{\bar{X}} = -.04$

Off-target difference $\quad D = M - \bar{\bar{X}} = 0 - (-.04) = .04$

Note that if D is negative, ignore the negative sign.

Process capability:

Process potential $\qquad\qquad C_p = \dfrac{W}{NT} = \dfrac{12}{9.66} = 1.23$

Upper process capability $\quad\text{CPU} = \dfrac{2(\text{USL} - \bar{\bar{X}})}{\text{PS}} = \dfrac{2[6 - (-.04)]}{9.66} = 1.25$

Lower process capability $\quad\text{CPL} = \dfrac{2(\bar{\bar{X}} - \text{LSL})}{\text{PS}} = \dfrac{2[-.04 - (-6)]}{9.66} = 1.23$

Location index $\qquad\qquad K = \dfrac{2D}{W} = \dfrac{2 \times .04}{12} = .007$

Process capability $\qquad\quad C_{pk} = C_p(1 - k) = 1.23(1 - .007) = 1.22$

Note that for a unilateral specification C_{pk} equals either CPU or CPL.

7.4 From Chart 3.2 $\bar{\bar{X}} = 0.46$ and $\bar{R} = 7.5$ with a reasonably stable process. The capability indices for specification limits of ± 13 are computed below.

$$\hat{\sigma} = \frac{\bar{R}}{d_2} = \frac{7.5}{2.33} = 3.2$$

$$C_p = \frac{\text{USL} - \text{LSL}}{6\hat{\sigma}} = \frac{13 - (-13)}{6 \times 3.2} = 1.35$$

$$\text{CPU} = \frac{\text{USL} - \bar{\bar{X}}}{3\hat{\sigma}} = \frac{13 - .46}{3 \times 3.2} = 1.31$$

$$\text{CPL} = \frac{\bar{\bar{X}} - \text{LSL}}{3\hat{\sigma}} = \frac{.46 - (-13)}{3 \times 3.2} = 1.40$$

$$k = \frac{2|0 - .46|}{13 - (-13)} = .04$$

$$C_{pk} = \text{minimum (CPL, CPU)} = 1.31$$

The process is approximately centered and now has reasonable capability.

Using the histogram data of the individual values gives $s = 3.5$, which gives $C_p = 1.24$ and $C_{pk} = 1.19$. These results are close to the control chart calculation even though the histogram in Problem 5.1 is not bell shaped.

7.5 From the control chart $\bar{R} = 1.6$, so $\hat{\sigma} = \bar{R}/d_2 = 1.6/2.33 = .69$. The process potential for LSL = 9, USL = 21 is

$$C_p = \frac{21 - 9}{6 \times .69} = 2.90$$

The mean is $\bar{\bar{X}} = 15.2$ (subgroup 5 is omitted from calculations), so the process centering index is

$$k = \frac{2|15 - 15.2|}{12} = .03$$

The process potential is

$$C_{pk} = C_p(1 - k)$$
$$= 2.90 \times .97$$
$$= 2.81$$

Using the histogram data gives $s = .88$, which gives $C_p = 2.27$ and $C_{pk} \doteq 2.20$. The increase in the estimated standard deviation is due to there being more values in the tails of the histogram than would normally be expected.

7.6 From Chart 7.4, $\bar{\bar{X}} = 22.1$, $\bar{R} = 14.4$, and the process is stable. The one-sided in-process specification is .003 inch (coded as 30), so the capability is

$$\hat{\sigma} = \frac{\bar{R}}{d_2} = \frac{14.4}{2.85} = 5.1$$

$$\text{CPU} = \frac{\text{USL} - \bar{\bar{X}}}{3\hat{\sigma}} = \frac{30 - 22.1}{3 \times 5.1} = .52$$

Clearly, the machine is not capable. This machine was replaced by a new machine whose tryout is discussed in the Chapter 7 case study.

7.7 (a and b) The means, standard deviations, and capabilities are as follows:

Machine	$\bar{X}$	s	C_p	C_{pk}
1	9.7	1.8	1.67	.87
2	12.8	2.2	1.36	1.18
3	18.0	1.8	1.67	.93
Combined	13.5	3.96	.76	.72

(c) Centering machines should result in process performance C_{pk} equal to potential C_p. Thus, $C_{pk} = 1.4$ (machine 2)–1.7 (machine 1 and 3) could be expected.

(d) Differences in the mean levels of stratification factors makes the process output more variable than if a common mean were present. Unfortunately, in many situations, unlike this problem, the stratification factors are initially unknown. They must be determined by collecting data for factors that may differ.

7.8 (a) The 20 consecutive parts give only a "snapshot" of the operation. This study will indicate the highest capability but not assess the required process stability. If no problem is detected using this study design, the operation may have stability problems that were not evaluated. However, if problems are detected (as in this study) this approach provides a quick method to determine if an obvious problem exists.

(b) The summary values are shown below:

	Spindle 1			Spindle 2		
	Minimum	Maximum	Difference	Minimum	Maximum	Difference
Mean	4.13	5.28	1.15	.05	1.55	1.5
SD	1.35	1.26	.33	1.07	1.23	.51
C_p	.74	.79		.93	.81	
CPL	0			.95		
CPU		0			.39	

The capability calculations assume stability of the individual spindles. There is an obvious difference in the mean levels for the two spindles. Spindle 1 is adjusted so its mean is beyond the specification limit. Process potential C_p is about the same for both spindles and shows that even if the spindle means were 0, capability would not be attained. Thus, centering and a reduction in variability are required.

(c) On the average, within-part variation is about 1 to 2 units out of a total specification width of 6 units. This difference will make attaining a capable process more difficult. A decision must be made on what diameter measurements are needed.

Solutions to the Problems in Chapter 9

9.1 (a, b, and c) The Pareto analysis work sheet is as follows:

Defect	Number of defects	% Composition	Cumulative percentage	% Total inspected
1. Poor cleanup	79	44	44	6.6
2. Dented molding	44	24	68	3.7
3. Mislocated side molding	24	13	81	2.0
4. Loose moldings	14	8	89	1.2
5. Scratched glass	7	4	93	0.6
6. Mislocated ball studs	5	3	96	0.4
7. Mislocated center molding	4	2	98	0.3
8. Other	3	2	100	0.3
Total	180	100	100	15.0

Figure 9.24 has Pareto diagrams for each of the three measures.

(d) The top three rejects account for 81% of the defects. A number of problems are related to moldings. The molding process and design should be evaluated for production robustness.

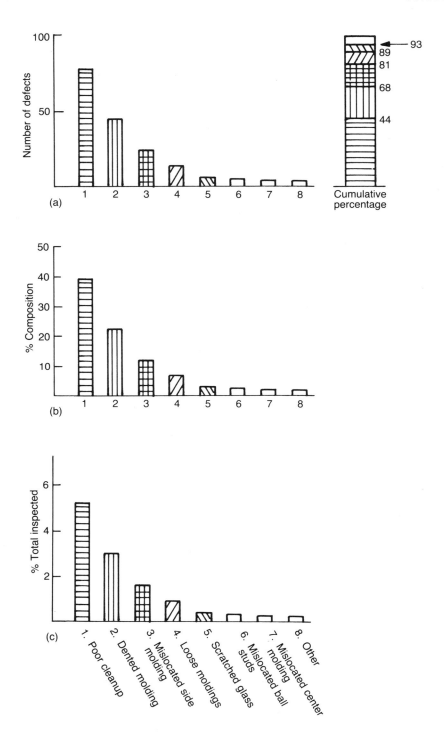

Figure 9.24 Pareto diagrams for Problem 9.1a–c: (a) number of defects; (b) percentage composition; (c) percentage of total inspected.

(e) The Pareto analysis cost work sheets appear as follows:

Defect	Number of defects	Cost per unit ($)	Cost of defect ($)	Rank
1. Poor cleanup	79	.30	23.70	3
2. Dented molding	44	.75	33.00	2
3. Mislocated side molding	24	.10	2.40	5
4. Loose molding	14	.25	3.50	4
5. Scratched glass	7	5.25	36.75	1
6. Mislocated ball studs	5	.35	1.75	6
7. Mislocated center molding	4	.16	.60	7
Total	180		101.70	

Defect	Cost of defect ($)	% Composition	Cumulative percentage
5. Scratched glass	36.75	36	36
2. Dented molding	33.00	33	69
1. Poor cleanup	23.70	23	92
4. Loose molding	3.50	3	95
3. Mislocated side molding	2.40	2	97
6. Mislocated ball studs	1.75	2	99
7. Mislocated center molding	.60	1	100

Figure 9.25 shows the cost Pareto diagram; it is apparent that three defects account for 92% of the total cost for defects. Interestingly, scratched glass increases dramatically in the prioritization of the problem due to the need to scrap rather than rework the tailgate glass.

An improvement team clearly should address the following problems:

1. Poor cleanup
2. Dented molding
3. Scratched glass

Since defect 3, mislocated side molding, is a different type of problem (i.e., mislocation), it can await resolution of these three problems.

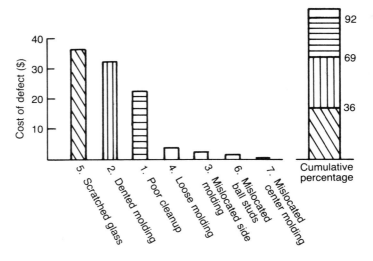

Figure 9.25   Cost Pareto diagram for Problem 9.1e.

9.2   The number of percentage of defects for the 11 days are as follows:

					Days							
Defect	1	2	3	4	5	6	7	8	9	10	11	Total
Loose vinyl	10	6	8	3	9	11	19	7	17	16	13	119
	1.8	1.2	2.2	.7	2.1	2.7	4.5	2.0	4.2	3.7	2.8	2.5
Total defects	38	24	25	28	23	36	37	20	28	29	22	310
	6.7	4.8	6.8	6.5	5.4	9.0	8.8	5.7	7.0	6.8	4.8	6.5

From Procedure 3.5 the control limit calculations using an average sample size $\bar{n} = 432$ are shown below.

Loose vinyl:

$$A_0 = 3\sqrt{\frac{2.5 \times 97.5}{432}} = 2.3$$

$$\text{UCL}_P = 2.5 + 2.3 = 4.8$$

$$\text{LCL}_P = 2.5 - 2.3 = .2$$

Total defects:

$$A_0 = 3\sqrt{\frac{6.5 \times 93.5}{432}} = 3.6$$

$$\text{UCL}_P = 6.5 + 3.6 = 10.1$$

$$\text{LCL}_P = 6.5 - 3.6 = 2.9$$

The *P* charts appear in Figure 9.26, where the processes for generating defects appear stable. This is only a preliminary indication since defects for too few days were available. This apparent stability implies that there is a predictable common-cause system generating the rejects. Unless the team is able to make fundamental changes to the processing system, the current total reject level will remain about 6.5%. The problem analysis system in Chapter 13 can be used to develop improvement actions.

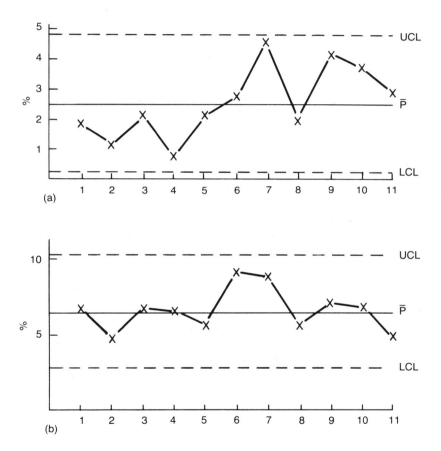

Figure 9.26   *P* charts for Problem 9.2: (a) loose vinyl defects; (b) total defects.

9.3 The Pareto analysis work sheet for Table 9.7 appears as follows:

Defect	Number of defects	% Composition	Cumulative percentage	% Total inspected
1. Distribution list	70	33	33	39.1
2. Job work sheet	29	14	47	16.2
3. Pagination	26	12	59	14.5
4. Title page	20	9	68	11.1
5. General format	15	7	75	8.4
6. Sequence	13	6	81	7.3
7. Document approval form	5	2	83	2.8
8. B/W prints	5	2	85	2.8
9. Drawing numbers	5	2	87	2.8
10. Table of contents	5	2	89	2.8
11. Reproduction card	4	2	91	2.2
12. Photomaster sheet	4	2	93	2.2
13. Caption	3	1	94	1.7
14. Others	8	4	98[a]	4.5
Total	212	98[a]		—[b]

[a]Note: does not total 100% due to rounding.
[b]Cannot be determined since number of reports having one or more defects M was not recorded.

An error that requires minimal time to correct is less serious, so a cost-type Pareto analysis should be prepared. The cost in this situation could be the average time required to correct an error. In most clerical applications this is an important consideration.

9.4 From Table 8.1 the rejects for the three highest frequency bores having rejects are by clock position:

Clock position	Bore		
	47	44	49
3	6 (3)	18 (14)	2 (8)
6	0 (0)	0 (0)	0 (0)
9	143 (79)	53 (42)	7 (29)
12	32 (18)	55 (44)	15 (63)
Total	181	126	24

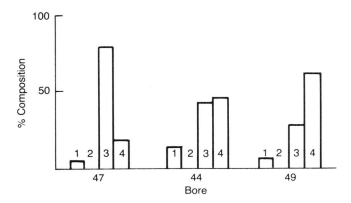

Figure 9.27   Pareto analysis for Problem 9.4.

Comparing a clock position across all bores is not meaningful in this situation because of the different orientation of the bores. Using percentage composition (in parentheses) enables evaluation of the importance of clock position. If clock position is not important, approximately equal percentages (25%) would appear in all positions. Figure 9.27 clearly indicates position is as important in understanding the cause of the rejects.

9.5   There are two logical explanations for both undersize and oversize ID problems.

Explanation 1, lack of machine capability:

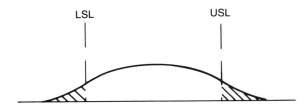

There is a general lack of capability for the entire machine.

Explanation 2, lack of common spindle distributions (stratification):

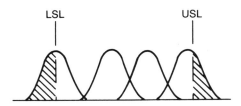

The spindles differ in mean and/or variability, causing rejects for both U/S and O/S. This could be investigated using a check sheet for defects that indicates spindle and any system differences in process flow. A faster method would be to conduct a spindle comparison study (see Fig. 11.6).

9.6 Since production rates varied, it is necessary to use the percentage of defects considering the actual parts produced:

Machine	Spindle	August	September	October	Total
A	1	5.0	4.3	5.3	4.9
	2	7.3	16.4	23.4	15.2
	Total	6.1	11.3	12.0	9.8
B	1	5.7	8.2	8.8	7.6
	2	33.5	15.5	6.0	16.2
	Total	16.8	13.5	7.6	12.3

(a) The percentage of defects for machines A and B during the 3 month period was 9.8 and 12.8%, respectively. The similar rates imply there is a common-cause system problem resulting in the high reject rate. Further stratification would clearly be useful.

(b) Figure 9.28a shows the machine by monthly Pareto analysis. The large variation in the percentage of defects implies special-cause problems could be influencing the output. A weekly P chart analysis would be appropriate to evaluate stability.

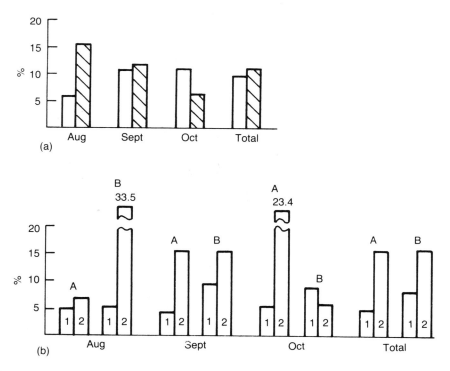

Figure 9.28 Pareto diagrams for Problem 9.6: (a) two-factor analysis; (b) three-factor analysis.

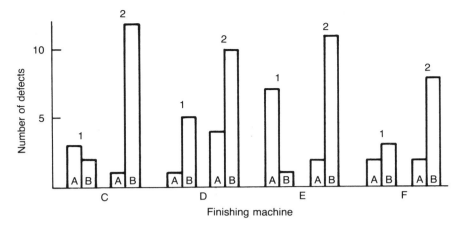

Figure 9.29   Pareto analysis for Problem 9.7: suppliers 1 and 2 and roughing machines A and B.

(c) Figure 9.28b shows the three-factor analysis. It is apparent that a special-cause problem exists for spindle 2. Either a design or setup problem may be present. Since spindle 2 is similar to spindle 1 during some periods, it is reasonable to conclude that spindle 2 must be more difficult to set up (or maintain) than spindle 1. About 5% of the defects appear to be a common-cause problem and the remainder a special-cause spindle 2 problem.

9.7   (a) The relatively constant rate of rejects across all finishing machines implies that a common-cause problem exists. Something common to all machines appears to be the cause of the increase in rejects. Since the number of rejects increased suddenly, something changed. Two possibilities exist: either upstream processing or processing on all four machines. The data collected by the team will help evaluate upstream factors. The team should also evaluate any changes to the finishing operation, such as PM procedures, tools, and common coolants, that would be common to all machines.

(b) Since the production rates differ, the data should generally be converted to percentages. For this problem, the number of units produced does not differ markedly, so use of the number of defects is possible—and easier. The Pareto diagram appears in Figure 9.29, where it is apparent that machine B has a problem processing material from supplier 2. The team must now address what changed in this part of the process.

Solutions to the Problems in Chapter 11

11.4 (a) Fixture variability differences can be evaluated using

$$F_{max} \text{ ratio} = \left(\frac{6.0}{3.5}\right)^2 = 2.9$$

The tabled value in Procedure 11.1 is 8.9, which is appreciably greater than 2.9, so we must conclude that there are no apparent differences in fixture variability. Generally, larger sample sizes (e.g., $n = 25$ or 30) are required to detect smaller (but important) differences commonly encountered in manufacturing. With $n = 10$, the F_{max} ratio would need to exceed 8.9 in this problem. Routine checks would likely have detected an obvious problem with a fixture if the difference were that large.

The pooled standard deviation is

$$s_p = \sqrt{\frac{(4.9)^2 + (5.0)^2 + (3.5)^2 + (3.7)^2 + (4.6)^2 + (4.6)^2 + (6.0)^2 + (4.4)^2}{8}}$$

$$= 4.6$$

which has $f = 9 \times 8 = 72$ degrees of freedom.

(b) Since there is common variability between fixtures, the ANOM procedure can be used. The grand mean is

$$\bar{\bar{X}} = \frac{16.3 + 18.0 + 19.8 + 20.0 + 18.0 + 19.4 + 19.5 + 23.2}{8}$$

$$= 19.3$$

The decision lines are

$$\text{UDL} = 19.3 + \frac{4.6 \times 2.8}{\sqrt{10}} \sqrt{\frac{7}{8}}$$

$$= 19.3 + 3.8$$

$$= 23.1$$

$$\text{LDL} = 19.3 - 3.8$$

$$= 15.5$$

The ANOM chart is shown in Figure 11.20. Fixture 8 appears to have higher flatness than the remaining fixtures.

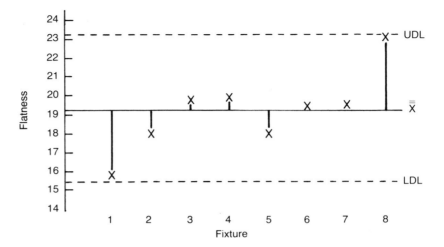

Figure 11.20 ANOM chart for flatness (Prob. 11.4).

11.5 (a) The mean and standard deviations are given in the solution to Problem 4.1b. Spindle variability differences can be evaluated using

$$F_{max} \text{ ratio} = \left(\frac{1.4}{.7}\right)^2 = 4.0$$

The tabular value in Procedure 11.1 is about 3.5, so there appear to be differences in spindle variability. Location differences will be evaluated using an LC chart.

(b) The values for the VC chart use $C_1 = .78$, $C_2 = 1.39$:

Spindle	s	UVL	LVL
1	1.15	1.60	.90
2	1.25	1.74	.98
3	1.09	1.52	.85
‘4	1.26	1.75	.98
5	.91	1.26	.71
6	1.14	1.58	.89
7	.66	.92	.51
8	1.40	1.95	1.09

The VC chart in Figure 11.21a shows that spindle 7 is less variable than most of the other spindles. Further investigation found that spindle 7 had a new collet recently installed. Spindles 1–5 had rebuilt collets installed. The new collets were judged to be a significant improvement over the rebuilt collets. The values for the LC chart use

$$\text{ULL} = \bar{X} + \frac{st}{\sqrt{n}} = \bar{X} + \frac{s2.06}{\sqrt{24}}$$
$$= \bar{X} + s.42$$

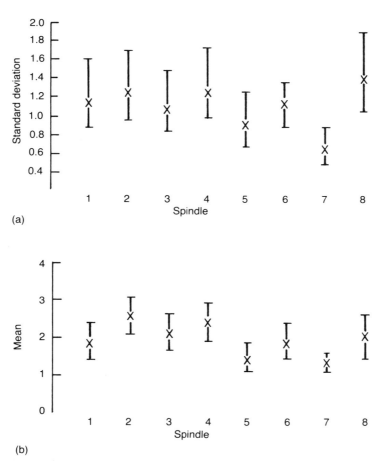

Figure 11.21 Comparison chart for eighth-spindle screw machine: (a) VC chart; (b) LC chart.

$$\text{LLL} = \bar{X} - \frac{st}{\sqrt{n}} = \bar{X} - s.42$$

Spindle	$\bar{X}$	s	ULL	LLL
1	1.90	1.15	2.38	1.42
2	2.55	1.25	3.08	2.02
3	2.17	1.09	2.63	1.71
4	2.41	1.26	2.94	1.88
5	1.43	.91	1.81	1.05
6	1.87	1.14	2.35	1.39
7	1.30	.66	1.58	1.02
8	2.05	1.40	2.64	1.46

The LC chart in Figure 11.21b shows that the new collet on spindle 7 also lowered the concentricity.

11.6 (a) The control chart values are, for the mean, $\bar{\bar{X}} = 11.1$, UCL = 12.6, LCL = 9.6, and for the range, $\bar{R} = 4.17$, UCL = 7.8, LCL = 0.6. The range chart is reasonably stable, and the $\bar{X}$ chart shows numerous out-of-control points.

    (b) The means and standard deviations are as follows:

Spindle	$\bar{X}$	s
1	10.4	2.7
2	11.2	1.8
3	12.4	1.8
4	11.8	2.6
5	11.6	1.9
6	10.7	2.7
7	10.8	2.2
8	9.8	2.0

Using the F_{max} ratio to compare spindle variability gives

$$F_{max} \text{ ratio} = \left(\frac{2.7}{1.8}\right)^2 = 2.3$$

which is well below the tabled value in Procedure 11.1. The common pooled standard deviation is

$$s_p = \sqrt{\frac{(2.7)^2 + (1.8)^2 + (1.8)^2 + (2.6)^2 + (1.9)^2 + (2.7)^2 + (2.2)^2 + (2.0)^2}{8}}$$

$$= 2.2$$

with $f = 184$ degrees of freedom. The limits for the ANOM chart are

$$\text{UDL} = 11.1 + \frac{2.2 \times 2.72}{\sqrt{24}}\sqrt{\frac{7}{8}}$$

$$= 11.1 + 1.1$$

$$= 12.2$$

$$\text{LDL} = 11.1 - 1.1$$

$$= 10.0$$

Fixtures 3 and 8 are beyond the decision lines. The LC chart limits are as follows:

Spindle	$\bar{X}$	LLL	ULL
1	10.4	9.5	11.3
2	11.2	10.3	12.1
3	12.4	11.5	13.3
4	11.8	10.9	12.7
5	11.6	10.7	12.5
6	10.7	9.8	11.6
7	10.8	9.9	11.7
8	9.8	8.9	10.7

The LC chart, shown in Figure 11.22, indicates that spindles 3 and 4 are significantly higher than spindle 8. The lack of stability on the $\bar{X}$ chart may be in part due to these spindle differences. In this case, a lack of stability and spindle stratification does not result in parts beyond the specification limits.

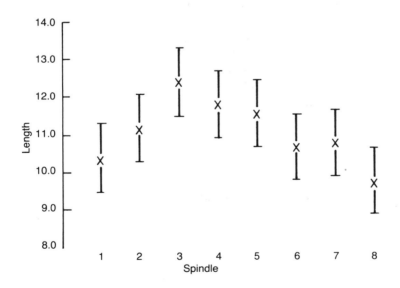

Figure 11.22 LC chart for gear length (Prob. 11.6).

11.7 (a) The control chart values for the $\bar{X}$ chart are $\bar{\bar{X}} = -.6$, UCL = 4.0, LCL = −5.2; for the R chart the values are $\bar{R} = 6.3$, UCL = 14.4. The range chart shows subgroup 9 with $R = 15$, which is beyond the control limit. The $\bar{X}$ chart shows no values beyond the control limits. However, there are no points in zone A, which with 50 subgroups is unlikely. This ''signal'' suggests that stratification (i.e., spindle differences) may exist.

(b) The means and standard deviations for the $n = 50$ values are as follows:

		Spindle		
	1	2	3	4
$\bar{X}$	1.5	$-.5$	-3.0	$-.4$
s	2.4	1.9	1.4	3.7

(c) The combined histogram in Figure 11.23 shows an unusual shape that results from spindle stratification.

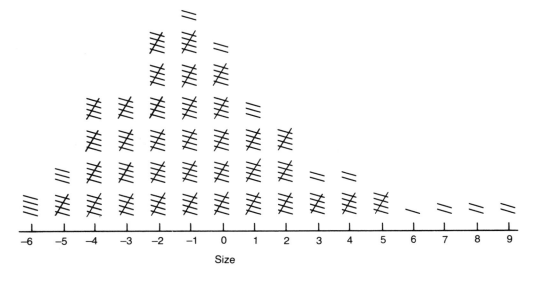

Figure 11.23   Histogram of journal size (Prob. 11.7).

(d) The F_{max} ratio for comparing variability is

$$F_{max} \text{ ratio} = \left(\frac{3.7}{1.4}\right)^2 = 6.98$$

which is greater than the tabled value in Procedure 11.1 of about 2.2. Thus, we conclude that there are variability differences between the spindles.

The VC chart values can be obtained using $C_1 = .83$, $C_2 = 1.25$:

		Spindle		
	1	2	3	4
UVL	3.0	2.4	1.8	4.6
LVL	2.0	1.6	1.2	3.1

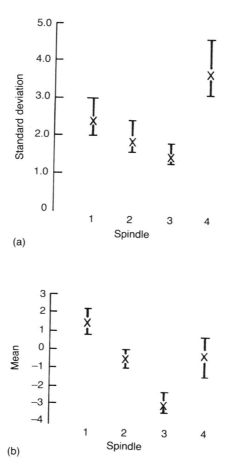

(a)

(b)

Figure 11.24   Comparison chart for four-spindle machine (Prob. 11.9): (a) VC chart; (b) LC chart.

The VC chart in Figure 11.24a shows that spindle 4 is clearly more viable than the remaining spindles. Also, spindle 3 appears to have lower variability than spindle 1.

   The LC chart values are as follows:

	Spindle			
	1	2	3	4
ULL	2.2	.0	−2.6	.7
LLL	.8	−1.0	−3.4	−1.5

It is apparent in Figure 11.24b that spindles 1 and 3 differ in location from spindles 2 and 4, which do not differ significantly from the 0 target.

(e) Ignoring stratification factors, such as spindles, make detection of large differences in location or variability difficult using control charts or histograms. As Procedure 4.1 discusses, it is necessary to eliminate any large differences using a baseline study. Comparison plots provide a useful tool in baseline studies.

11.8 (a) The study was designed so that all outside factors were balanced for the two loading methods. Thus, any difference will be due to the loading method.

(b) The positions cannot be compared using the methods in this chapter. Since the positions are measured on the same parts, the data are paired and Procedure 12.5 must be used. However, the summary values are as follows:

	Horizontal		Vertical	
	Position 1	Position 2	Position 1	Position 2
Mean	1.5	5.9	2.7	3.3
SD	1.4	2.2	1.8	1.4

and position 2 is appreciably higher than position 1 for horizontal loading.

(c) Procedure 11.4 should be used to compare variability and location. Variability differences are evaluated using, for position 1,

$$F_{max} \text{ ratio} = \left(\frac{1.8}{1.4}\right)^2 = 1.65$$

and for position 2,

$$F_{max} \text{ ratio} = \left(\frac{2.2}{1.4}\right)^2 = 2.47$$

The tabled value in Procedure 11.4 is 2.5. Thus, we conclude no variability differences are detected. However, since the position 2 ratio is close to 2.5, a larger sample size may detect a difference. Comparing the means results in

$$s_p = \sqrt{\frac{(1.4)^2 + (1.8)^2}{2}} \qquad \text{position 1}$$

$$= 1.6$$

$$s_p = \sqrt{\frac{(2.2)^2 + (1.4)^2}{2}} \qquad \text{position 2}$$

$$= 1.8$$

$$f = 15 + 15 - 2$$

$$= 28$$

$$a = \sqrt{\frac{15 + 15}{15 \times 15}}$$

$$= .37$$

$$T = \frac{2.7 - 1.5}{.37 \times 1.6} \qquad \text{position 1}$$

$$= 2.02$$

$$T = \frac{5.9 - 3.3}{.37 \times 1.8} \qquad \text{position 2}$$

$$= 3.90$$

The tabled value in Procedure 11.4 is about 2.04. Thus, the vertical method is a significant improvement for position 2. There may be some deterioration in position 1 using vertical loading, but the overall reduction in roundness strongly favors this method. Since position 2 is more out of round than position 1, improvement should focus on position 2.

11.9 (a) The summary values for the two methods are as follows:

	Current process	New process
n	25	15
$\bar{X}$	4.49	5.37
s	1.58	.97

Variability differences are evaluated using

$$F_{\max} \text{ ratio} = \left(\frac{1.58}{.97}\right)^2 = 2.65$$

The tabled value in Procedure 11.4 corresponding to 25 and 15 is 2.4, so we conclude that the processes have differing variability.

Location differences are evaluated by the following computations:

$$w_1 = \frac{(1.58)^2}{25} = .10$$

$$w_2 = \frac{(.97)^2}{15} = .063$$

$$w = \sqrt{.10 + .063} = .40$$

$$T = \frac{5.37 - 4.49}{.40} = 2.2$$

The degrees of freedom for T are calculated as follows:

$$u_1 = \frac{(.10)^2}{24} = .00042$$

$$u_2 = \frac{(.063)^2}{14} = .00028$$

$$f = \frac{(.163)^2}{.00042 + .00028} = 37$$

The tabled t value is about 2.03, so we conclude that the means are significantly different for the two processes. The new process is superior since it has higher yield and lower variability.

(b) Calculations for the comparison charts are shown below.

VC chart:

$$UVL = 1.58 \times 1.39$$
$$= 2.20$$
$$\qquad\qquad \text{current process}$$
$$LVL = 1.58 \times .78$$
$$= 1.23$$
$$UVL = .97 \times 1.58$$
$$= 1.53$$
$$\qquad\qquad \text{new process}$$
$$LVL = .97 \times .73$$
$$= .71$$

LC chart:

$$ULL = 4.49 + \frac{1.58 \times 2.06}{\sqrt{25}}$$
$$= 4.49 + .65$$
$$= 5.14 \qquad\qquad \text{current process}$$
$$LLL = 4.49 - .65$$
$$= 3.84$$
$$ULL = 5.37 + \frac{0.97 \times 2.14}{\sqrt{15}}$$
$$= 5.37 + .54$$
$$= 5.91 \qquad\qquad \text{new process}$$
$$LLL = 5.37 - .54$$
$$= 4.83$$

11.10   (a) The summary values for the two groups are as follows:

	Nonheat treated	Heat treated
n	25	50
$\bar{\bar{X}}$	1.26	.59
s	3.11	2.17

Variability differences are evaluated using

$$F_{max} \text{ ratio } = \left(\frac{3.11}{2.17}\right)^2 = 2.05$$

The tabular value in Procedure 11.4 corresponding to 25 and 50 is 1.7, so we conclude that the heat treatment reduces variability.

Location differences are evaluated by the following computations:

$$w_1 = \frac{(3.11)^2}{25} = .39$$

$$w_2 = \frac{(2.17)^2}{50} = .094$$

$$w = \sqrt{.39 + .094} = .70$$

$$T = \frac{1.26 - .59}{.70} = .96$$

The degrees of freedom for T are calculated as follows:

$$u_1 = \frac{(.39)^2}{24} = .0063$$

$$u_2 = \frac{(.094)^2}{49} = .00018$$

$$f = \frac{(.484)^2}{.0063 + .00018} = 36$$

The tabular t value is about 2.03, so we conclude that the means are not significantly different.

(b) Calculations for the comparison charts are shown below.

VC charts:

$$\text{UVL} = 3.11 \times 1.39$$
$$= 4.32$$

not heat treated

$$\text{LVL} = 3.11 \times .78$$
$$= 2.42$$

$$\text{UVL} = 2.17 \times 1.25$$
$$= 2.71$$

heat treated

$$\text{LVL} = 2.17 \times .83$$
$$= 1.80$$

LC charts:

$$\text{ULL} = 1.26 + \frac{3.11 \times 2.06}{\sqrt{25}}$$

$$= 1.26 + 1.28$$

$$= 2.54 \qquad \text{not heat treated}$$

$$\text{LLL} = 1.26 - 1.28$$

$$= -.02$$

$$\text{ULL} = .59 + \frac{2.17 \times 2.01}{\sqrt{50}}$$

$$= .59 + .62$$

$$= 1.21 \qquad \text{heat treated}$$

$$\text{LLL} = .59 - .62$$

$$= -.03$$

Solutions to the Problems in Chapter 12

12.1 (a) The two scatter plots appear in Figure 12.50. There appears to be no correlation between the two diameters. Since the data are coded, the benchmark line indicates identical points within the specification range of the two diameters. Both plots have most points above the benchmark line, indicating that the small diameters are above the large diameters within their respective specification intervals.

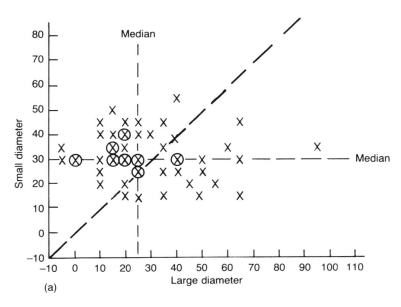

Figure 12.50 Scatter plots of large and small valve diameter: (a) before deburr; (b) after anodize.

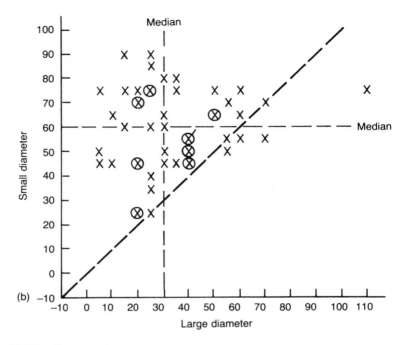

Figure 12.50 Continued.

(b) For the before deburr situation, the sign test results in

Quadrant	No. of points	
1	8	$A = 8 + 4 = 12$
2	11	$B = 11 + 8 = 19$
3	4	$C = \min(A, B) = 11$
4	8	$Q = 19$
Total	31	$SS = 50 - 19 = 31$

The tabled value in Procedure 12.2 is $c = 9$ for $SS = 31$, so we conclude that correlation does not exist. The correlation coefficient is $r = -.11$, which is below the critical value in Procedure 12.3.

For the after anodize situation, the sign test results in

Quadrant	No. of points	
1	10	$A = 10 + 10 = 20$
2	11	$B = 11 + 12 = 23$
3	10	$C = \min(A, B) = 20$
4	12	$Q = 7$
Total	43	$SS = 50 - 7 = 43$

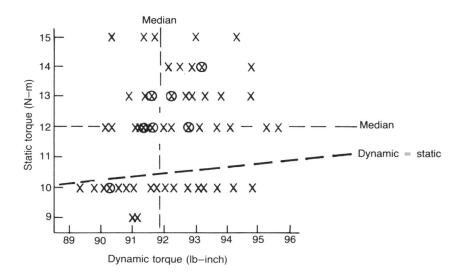

Figure 12.51 Scatter plot of dynamic versus static torque.

The tabled value in Procedure 12.2 is $c = 14$ for SS = 43, so we conclude that correlation does not exist. The correlation coefficient is $r = .14$, which is below the critical value in Procedure 12.3.

(c) Since the large-diameter before deburr and after anodize measurements are highly correlated, as seen in Figure 12.41a, there is little gain in having two control charts. Upstream control is usually desirable, so a chart before deburr is adequate. Unfortunately, the large and small diameters are not correlated, so a single large-diameter chart will not monitor the behavior of the small diameter. Also, before deburr and after anodize measurements are not related for the small diameter, as seen in Figure 12.41b. Thus, two charts may be required for the small diameter. Because of these differences in behavior between the large and small diameters, the small-diameter process should be improved so that greater predictability exists. Less monitoring will then be necessary.

12.2 (a) The scatter plot appears in Figure 12.51, where there appears to be little relationship between the dynamic and static torque. Since the scales are different, the benchmark line must be adjusted for the scale difference. Selecting two points 89 and 96 results in

89 (pounds-inch) $\times$ .113 = 10.06 (N-m)

96 (pounds-inch) $\times$ .113 = 10.85 (N-m)

The benchmark line can be drawn by connecting the points (89, 10.06) and (96, 10.85). It is apparent that there is no close correspondence between the two measurements.

(b) It is not necessary for the two torque readings to be equal to enable effective monitoring. However, a high correlation is required so that a meaningful prediction line can be computed:

Static torque $= a + b$ (dynamic torque)

Using the sign test for correlation results in

Quadrant	No. of points	
1	15	$A = 15 + 14 = 29$
2	7	$B = 7 + 6 = 13$
3	14	$C = \min(A, B) = 13$
4	6	$Q = 18$
Total	42	$SS = 60 - 18 = 42$

The tabled value in Procedure 12.2 is $c = 14$, so we conclude that correlation does exist. The correlation coefficient is $r = 0.30$, which is significant from Procedure 12.3.

(c) If the two measurements do not have a high correlation, setup and monitoring using dynamic torque are not meaningful. It may be necessary to determine the bolt's functional requirement in terms of dynamic torque so that process performance can be effectively evaluated.

(d) The histograms for the two torque readings appear as follows:

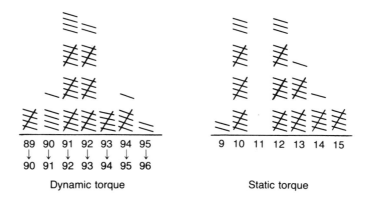

Dynamic torque   Static torque

It is apparent that a problem exists with the static torque measurements since no values of 11 N-m are observed. Measurement variability increases the scatter of the data, reducing the appearance of any correlation. Also, the lack of precision of the static torque measurement scale further contributes to an apparent low correlation.

12.3 (a) The scatter plot appears in Figure 12.52a. An apparent correlation exists between OAL and thrust face size.

(b) The correlation coefficient is $r = -.75$, which is above the critical value in Procedure 12.3.

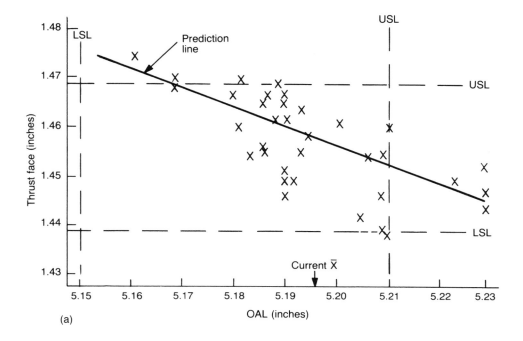

(a)

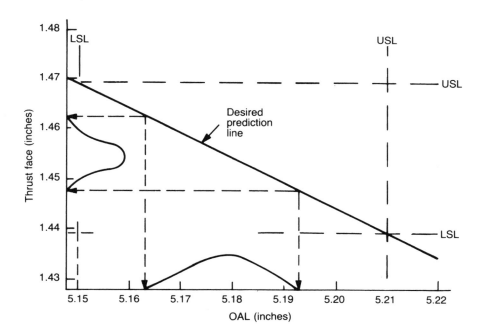

Figure 12.52   Relationship of gear blank OAL to thrust face slip: (a) actual and (b) desired relationship.

(c) The prediction line is

Thrust face $= 3.590 - .411$ OAL

To draw the prediction line on the scatter plot, select two OAL points, say 5.16 and 5.22:

$3.590 - (.411 \times 5.16) = 1.469$

$3.590 - (.411 \times 5.22) = 1.445$

The prediction line can be drawn by connecting the points (5.16, 1.469) and (5.22, 1.445).

(d) Retargeting OAL would not produce a significant improvement. Increasing OAL would make OAL values beyond the upper specification limit more likely. Reducing OAL would make thrust face sizes too large. The best strategy for improvement would be to reduce OAL variability by reducing gram weight variability. Also, changing the relationship between OAL and thrust face size would have a significant benefit.

(e) Ideally, the prediction line would connect the corners of the "specification box" shown in Figure 12.52b. This line connects the points (5.15, 1.469) and (5.21, 1.439). This line enables maximum use of the specification band for both characteristics. The nominal OAL is 5.18, which from the prediction line determined in the case study results in

$$\text{Weight} = \frac{\text{OAL} + 2.54}{.0082} = 941.5 \text{ g}$$

12.4   (a) The scatter plot appears in Figure 12.53. Burnishing improves the surface finish. The plot also indicates that there is a relationship between surface finish before and after burnishing.

(b) It is possible to reduce the mean without the existence of correlation. The sign test for correlation results in

Quadrant	No. of points	
1	12	$A = 12 + 11 = 23$
2	0	$B = 0 + 0 = 0$
3	11	$C = \min(A, B) = 0$
4	0	$Q = 7$
Total	23	$SS = 30 - 7 = 23$

The tabled value in Procedure 12.2 is $c = 6$ for $SS = 23$, so we conclude that correlation exists. The correlation coefficient $r = .78$ is above the critical value in Procedure 12.3.

(c) The summary values are as follows:

	Before burnishing	After burnishing
Mean	35.6	15.5
SD	10.2	8.6

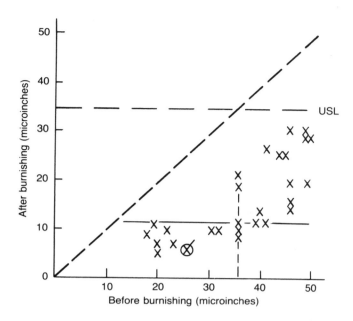

Figure 12.53   Scatter plot of surface finish before and after burnishing.

From Procedure 12.5,

$$F_{max} = \left(\frac{10.2}{8.6}\right)^2 = 1.41$$

$$J = (4)(.78)^2(1.41) = 3.43$$
$$K = (1.41 + 1)^2 = 5.81$$
$$L = 1.41 - 1 = .41$$
$$M = \frac{.41}{\sqrt{5.81 - 3.43}} = .27$$

which is below the .36 critical value. Thus, we conclude burnishing does not significantly reduce variability.

(d) The prediction line is

After burnishing $= -8.0 + .66$ (before burnishing)

Because of the high correlation, it is necessary to have a sufficiently low microfinish coming into the burnishing operation. Rearranging the equation and substituting the specification limit results in

$$\text{Before burnishing} = \frac{\text{after burnishing} + 8.0}{.66}$$

$$= \frac{35 + 8}{.66}$$

$$= 65$$

Because of variability around the prediction line, incoming values below about 50 are required. This type of result is significant since it demonstrates that final part quality depends directly on in-process quality.

12.5 (a) The histograms for both groups of before measurements uniformly span slightly beyond the 0–70 in-process specification interval. This uniform pattern is expected since the parts were specially selected. The after-finishing histograms show that machine 2 exhibits lower runout.

(b) It is not necessary to compare the before measurements using Procedure 11.4. Parts were randomly assigned to machines 1 and 2, so no differences would be expected. Also, the nonnormal histogram in part a makes use of Procedure 11.4 invalid. Since the before and after measurements are highly correlated, differences between the two groups of before measurements would influence the interpretation of the after comparison.

(c) The scatter plot appears in Figure 12.54. Data from both machines is below the benchmark line, indicating an improvement in runout. The before and after results are obviously correlated. Machine 1 results are generally above machine 2 results, indicating a stratification effect as was seen in the histograms.

(d) Using Procedure 12.5, the mean difference between the before and after measurements can be compared. However, the improvement is obvious. Variability cannot be compared using Procedure 12.5 because of the nonnormal sampling method. It is apparent that variability is reduced in the operation and that machine 2 has lower variability.

(e) The histograms of the machine 1 and 2 output do not represent a normal, bell-shaped distribution. Selecting parts that spanned the specification caused the nonnormal shape. The difference between machines 1 and 2 can be determined by comparing the change (after − before) due to each machine. Machine 1 had a change of -14.1 ($T = 8.2$) and machine 2 had -26.1 ($T = 7.2$). Thus, we conclude that machine 2 reduces runout by 12 more units than machine 1, on the average.

(f) The prediction line factors (a, b) can be computed from the summary values. For machine 1,

$$b = \frac{.97 \times 18.2}{24.9} = .71$$

$$a = 32.7 - (.71 \times 46.8) = -.53$$

The prediction lines for all machines are:

Machine 1:   after runout $= -.53 + .71$ (before runout)
Machine 2:   after runout $= 4.4 + .31$ (before runout)
Combined: after runout $= 1.6 + .51$ (before runout)

The three lines are drawn in Figure 12.54.

The difference between the two machines is apparent. The combined prediction line shows the line that would have been obtained if the data were not stratified by machine. Thus, excessive variability around a prediction line can be due to the presence of unaccounted stratification factors.

If this study had been performed using standard production parts in the 10–40 range, the difference between the machines would not have been as apparent. Spreading out the before parts highlighted the differences.

(g) The problem is that machine 1 does not reduce the runout as effectively as machine 2. When more extreme samples in the 40–70 range are encountered, machine 1 will not reduce the runout adequately. Improvement will involve making machine 1 perform like machine 2.

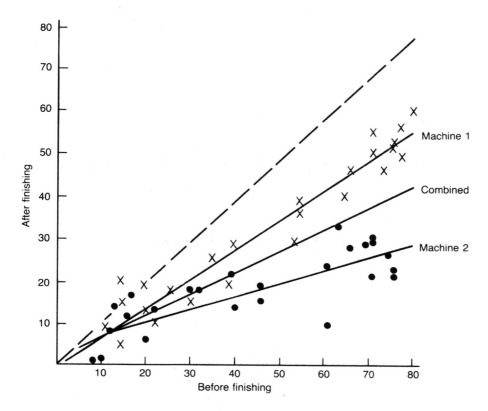

Figure 12.54   Runout before and after a finishing operation: machine 1 (×) and machine 2 (●).

Comparing the production output of both machines using a large enough sample size would likely have indicated machine 1 was a problem. However, the study approach that was used completely characterized the behavior of the finishing operation and the capability problem.

12.6   (a) The two scatter plots appear in Figure 12.55a and b. A strong correlation does not exist in either case. Since both groups of points are below the benchmark line, flatness does improve through the process. Interestingly, several casting flatness measurements were better than the finished measurements for a part. Since castings are difficult to measure, it is possible that measurement error influenced the results.

(b For the casting versus finished plot, the sign test results in

Quadrant	No. of points	
1	16	$A = 16 + 15 = 31$
2	7	$B = 7 + 7 = 14$
3	15	$C = \min(A, B) = 14$
4	7	$Q = 5$
Total	45	$SS = 50 - 5 = 45$

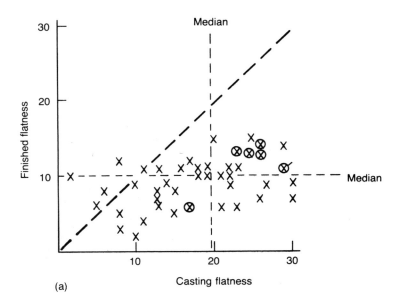

(a)

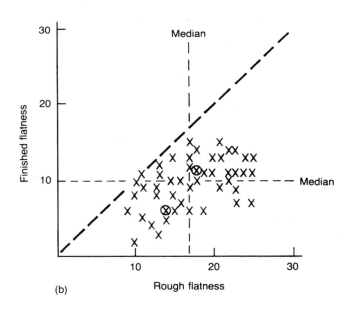

(b)

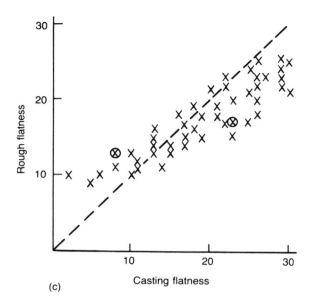

(c)

Figure 12.55 Scatter plots of flatness changes in a process; (a) casting versus finished; (b) rough versus finished; (c) casting versus rough.

The tabled value in Procedure 12.2 is $c = 15$ for SS = 45, so we conclude that correlation does exist. The correlation coefficient is $r = 0.49$, which is significant from Procedure 12.3.

For the rough part versus the finished part, the sign test results in

Quadrant	No. of points	
1	16	$A = 16 + 15 = 31$
2	4	$B = 4 + 5 = 9$
3	15	$C = \min(A, B) = 9$
4	5	$Q = 10$
Total	40	SS = 50 − 10 = 40

The tabled value in Procedure 12.2 is $c = 13$ for SS = 40, so we conclude that correlation does exist. The correlation coefficient is $r = 0.48$, which is significant from Procedure 12.3. Thus, in both cases we conclude that mild correlation exists.

(c) It is apparent in Figure 12.55c that a strong correlation exists ($r = 0.89$) between casting and rough part flatness.

(d) The in-process flatness relationships are as follows:

$$\text{Casting} \xrightarrow{\text{strong}} \text{roughing} \xrightarrow{\text{weak}} \text{finishing}$$
$$\xrightarrow{\text{weak}}$$

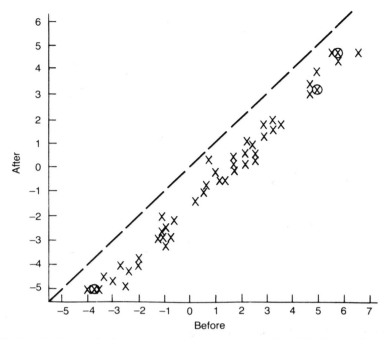

Figure 12.56   Before and after heat-treat measurements of a critical part dimension.

Thus, the finishing operation is not strongly influenced by incoming stock. This behavior is not necessarily bad. It implies that improvements to finished flatness should address the finishing operation. Additional potential causes should be developed in light of this conclusion.

12.7   (a) The scatter plot appears in Figure 12.56. A strong correlation obviously exists. The before measurements are about 1.5 units higher than the after measurements.

(b) The correlation coefficient is $r = 0.99$, which confirms the interpretation of the scatter plot.

(c) The data were collected for this study in a nonrandom manner. The spread of the data was purposely expanded to determine the influence of heat treat over the expected range of operation. The histograms for the before and after measurements confirm the expected nonnormal shape. Thus, it is not meaningful to use Procedure 12.5 to quantify differences in variability.

It is possible to evaluate a mean shift using Procedure 12.5 since the differences have an approximate normal distribution. The summary values are as follows:

	Before	After	Difference
Mean	1.0	−.4	−1.5
SD	3.0	3.1	.37

Since $T = 28$, there is an obvious mean shift.

(d) The prediction line factors are as follows:

$$b = \frac{.99 \times 3.1}{3.0} = 1.0$$

$$a = -.4 - (1.0 \times 1.0) = -1.4$$

Thus, the prediction line is

After heat treat $= -1.4 + 1.0$ (before heat treat)

(e) The prediction line can be used to determine the shift due to heat treat at various points within the specification range:

At -4:	after $= -1.4 + (1.0 \times -4) = -5.4$	change $= -1.4$
At 0:	after $= -1.4 + (1.0 \times 0) = -1.4$	change $= -1.4$
At 5:	after $= -1.4 + (1.0 \times 5) = 3.6$	change $= -1.4$

The slope of $b = 1.0$ implies that a constant change of $a = -1.4$ can be expected across the -4 to 6 interval. Prediction should not be made outside the interval where the data were collected.

The before heat-treat process should be targeted 1.4 units high to compensate for shrinkage during heat treat. Using the prediction line approach for targeting is preferable when strong correlation exists and it is possible that the target may change across the range of the before measurements. Examining the prediction line enables evaluation of this possibility.

(f) With the high correlation, only a before heat-treat control chart is required.

12.8 (a) The two scatter plots appear in Figure 12.57. There is no obvious relationship between the before and after heat-treat lead measurements. The specification is 0 to -5, and the observed parts span only about 40% of this range. Thus, the lack of correlation may be due to the narrow range of the before data. It is apparent that the two machines are targeted differently before heat treat.

(b) The correlation coefficients are $r = -.01$ for machine 1 and $r = -.14$ for machine 2. From Procedure 12.3, there is no correlation. A prediction line would not be meaningful.

(c) Using Procedure 11.4, the differences for the two machines can be compared. These differences are the shift due to the machine. From step 3,

$$F_{max} = \left(\frac{1.44}{1.00}\right)^2 = 2.07$$

The critical value is 1.6 from Procedure 11.4. Thus, we conclude machine 2 is more variable in its shift. From step 6,

$$w_1 = \frac{(1.00)^2}{47} = .021$$

$$w_2 = \frac{(1.44)^2}{47} = .044$$

$$w = \sqrt{.021 + .044} = .25$$

$$T = \frac{7.16 - 6.67}{.25} = 1.96$$

$$u_1 = \frac{(.021)^2}{46} = .0000097$$

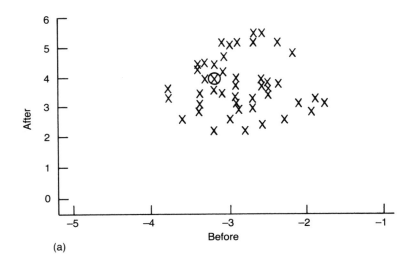

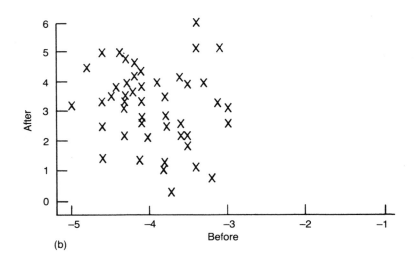

Figure 12.57   Gear lead before and after heat treat: (a) machine 1; (b) machine 2.

$$u_2 = \frac{(.044)^2}{46} = .000042$$

$$f = \frac{(.021 + .044)^2}{.000052} = 81$$

The critical value is 1.98, which is close to 1.96, so we conclude that there may be a difference in the mean shift of the two machines. However, the shift is so small that it is of little practical significance.

(d) The average shift due to heat treat is 6.9. Thus, a nominal of 0 after heat treat would be obtained by a −6.9 mean before heat treat. A worst case before spread would be obtained using the higher machine 2 standard deviation (.5). Thus, the upper and lower process limits for the before measurements are as follows:

Before UPL = $-6.9 + (3 \times .5) = -5.4$

Before LPL = $-6.9 - (3 \times .5) = -8.4$

An in-process specification of -5 to -9 would seem appropriate. Assuming a 0 after target is attained, since $C_p = 1.6$ minimal problems should be encountered.

(e) The variability before and after heat treat can be compared using Procedure 12.5. For machine 1,

$$F_{max} = \left(\frac{.87}{.48}\right)^2 = 3.29$$

$$J = (4)(-.01)^2(3.29) = .001$$

$$K = (3.29 + 1)^2 = 18.4$$

$$L = 3.29 - 1 = 2.29$$

$$M = \frac{2.29}{\sqrt{18.4 - .001}} = .53$$

which is significant. For machine 2 $M = .74$, which is also significant. Heat treat increases the variability of parts from both machines, but machine 2 has a much greater increase. Perhaps the internal stresses (see fourth case study) introduced by the machines differ.

(f) Several actions are required to improve the process:

The before heat-treat process should be targeted around -7.

Before heat-treat location stratification should be eliminated by constant targeting.

The cause for the increased after heat-treat variability from machine 2 should be eliminated.

12.9 (a) There appears little relationship between ID size or runout from Op 10 to gear lead from Op 20. The correlation coefficients are as follows:

Op 10 ID size with Op 20 gear lead, $r = -.05$.

Op 10 ID runout with Op 20 gear lead, $r = -.02$.

However, parts were randomly collected and did not span much of the specification range. At least, for the range that was studied no correlation exists. Since no relationship exists, the gear lead produced in Op 20 is insensitive to the incoming ID size or runout. This robustness is desirable for a process.

(b) The comparison of size and runout uses the following summary values:

	Op 20 ID size			Op 20 ID runout		
	Before	After	Difference	Before	After	Difference
Mean	2.19	$-.79$	-2.98	1.67	2.27	.60
SD	1.41	1.81	1.06	.91	.99	1.15
	$T = (\sqrt{50} \times 2.98)/1.06$			$T = (\sqrt{50} \times .60)/1.15$		
	$= 19.9$			$= 3.69$		

These results clearly suggest that the ID size, and to a lesser extent runout, change in Op 20. These quantitative results must be interpreted in light of process knowledge. In Op 20 the ID is used for clamping only. Differences in the ID may in fact be due to changes in ID roundness, which would make measurement of ID size subject to interpretation.

The ID size change in Op 30 can be evaluated similarly:

	Before	After	Difference
		Op 30 ID size	
Mean	$-.79$	$-.71$	.08
SD	.99	1.67	.88
	$T = (\sqrt{50} \times .08)/.88 = 0.64$		

Clearly, the mean size does not change in Op 30.

(c) An ANOM chart can be used to evaluate any stratification. The upper and lower decision lines can be computed using Procedure 11.3, with $h = 2.95$:

$$\frac{1.4 \times 2.95}{\sqrt{5}} \sqrt{\frac{9}{10}} = 1.75$$

$$\text{UDL} = 2.18 + 1.75 = 3.93$$

$$\text{LDL} = 2.18 - 1.75 = .43$$

The ANOM plot appears in Figure 12.58a. It is apparent that severe stratification exists. Lead rejects can be improved by targeting Op 20 machines consistently (see Prob. 12.8).

(d) Again an ANOM chart can be used to assess any runout stratification. The decision lines are as follows:

$$\frac{.87 \times 2.95}{\sqrt{5}} \sqrt{\frac{9}{10}} = 1.09$$

$$\text{UDL} = .6 + 1.09 = 1.69$$

$$\text{LDL} = .6 - 1.09 = -.49$$

The ANOM plot appears in Figure 12.58b. It is apparent that machine 10 significantly increases the ID runout more than the other machines. The machine stratification could contribute to a high reject rate.

12.10 (a) The correlation coefficients are given below:

	Hole 1	Hole 2	Hole 3	Hole 4
Hole 1	1.0			
Hole 2	0	1.0		
Hole 3	$-.84$	.11	1.0	
Hole 4	$-.30$	$-.83$	.22	1.0

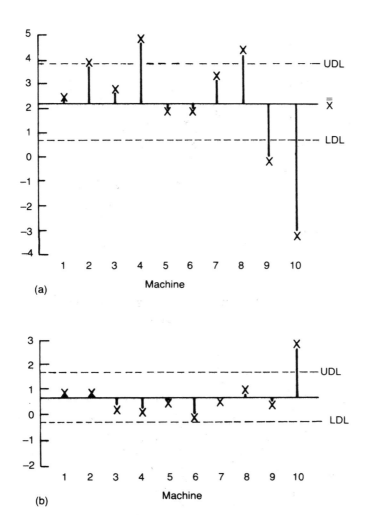

Figure 12.58 ANOM chart to elevate Op 20 hob stratification: (a) gear lead after Op 20; (b) change in ID runout in Op 20.

For $N = 32$, correlations below .33 are not significant. Thus, only hole pairs 1–3 and 2–4 are related.

(b) The scatter plot for the related pairs appears in Figure 12.59. An inverse relationship exists between the position of holes 1–3 and 2–4.

(c) If the end-cutting reamer completely determined the position of the holes, all holes should have similar correlations. Since hole pairs 1–3 and 2–4 are correlated, the obvious conclusion considering the part processing is that rough positioning of the holes is important. Several quality improvements on the rough stations did in fact dramatically improve final part quality. The end-cutting reamer did not "cut its own hole"!

12.11 (a) The scatter plot appears in Figure 12.60. Fixture and part runout are related.

(b) The correlation coefficient is $r = .93$. The summary values are as follows:

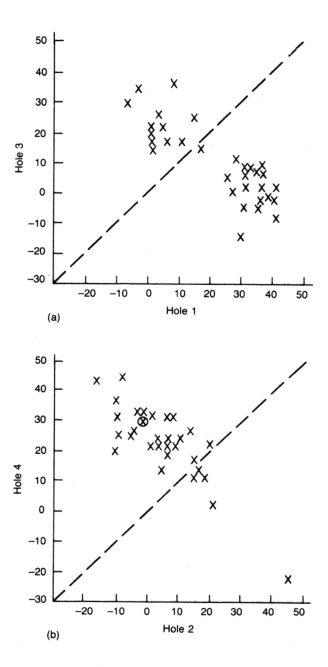

Figure 12.59   Hole true position.

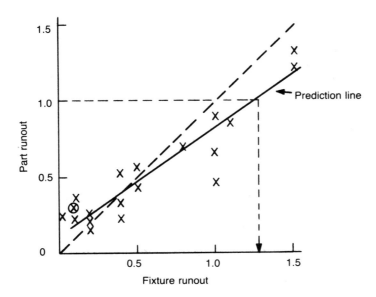

Figure 12.60 Scatter plot of fixture and part runout.

	Fixture	Part
Mean	.56	.52
SD	.48	.34

Thus the regression line factors are

$$b = \frac{.93 \times .34}{.48} = .66$$

$$a = .52 - (.66 \times .56) = .15$$

which gives the prediction line

Part runout $= .15 + .66$ (fixture runout)

(c) For a part runout of 1.0, the fixture runout is

$$\text{Fixture runout} = \frac{\text{part runout} (= 1.0) - .15}{.66} = 1.29$$

However, this predicted value does not account for variability around the prediction line. Also, it is apparent that many fixtures have runout below about .5. Why can't we make all fixtures have low runout and significantly improve part quality? A specification of about .7 should be achievable.

(d) Only one part was used from each fixture. Using more parts (say, five or more) from each fixture would reduce variability about the prediction line.

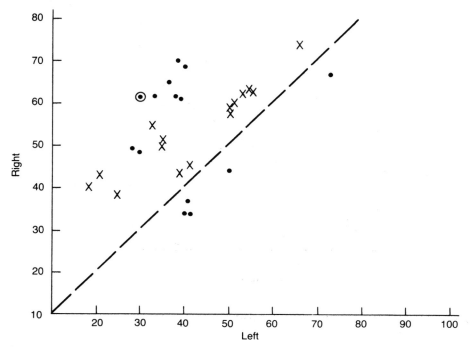

Figure 12.61 Comparison of right and left sides of a part for two fixtures: fixture 11 (✕) and fixture 12 (●).

12.12   (a) A shift in the mean implies nothing about correlation.

(b) The stratified scatter plot appears in Figure 12.61. A strong relationship between the right and left sides exists for fixture 11 but not for fixture 12.

(c) The correlation coefficients for the fixtures are as follows:

Fixture	Correlation
1	.56
2	.77
3	.30
4	.13
5	.62
6	.83
7	.62
8	.19
9	.64
10	.89
11	.91
12	.07
13	.82
14	.71
15	.77

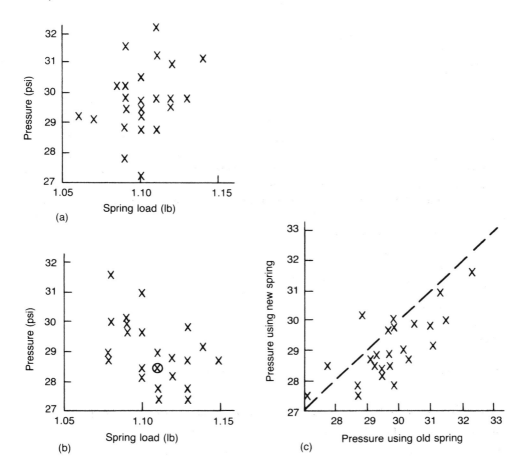

Figure 12.62 Evaluation of new spring design: (a) old spring; (b) new spring; (c) pressure comparison.

From Procedure 12.3, values above .51 are considered significant.

(d) Some fixtures produce high correlations between the right and left side dimensions, and some do not. This stratification suggests an inconsistency in the process that should be eliminated. Perhaps inconsistent positioning of a part within a fixture produces low correlations due to high variability. Mean shifts are not related to this problem.

12.13   (a) The scatter plots appear in Figure 12.62a and b. There does not exist a strong relationship. The correlation coefficients are as follows:

Old spring $r = .33$.
New spring $r = -.52$.

From Procedure 12.3, .40 is the critical value so we conclude that only the new spring exhibits any relationship between pressure and spring load. If engineering theory suggests a strong relationship should exist, we must question the measurement system for both pressure and spring load. A lack of measurement repeatability would introduce variability that would mask any correlation. Appendix I presents a method to assess repeatability.

(b) The summary values are as follows:

	Old spring	New spring
Mean	1.101	1.108
SD	.018	.020

From Procedure 11.4, $s_p = .019$, $a = .29$, and $T = 1.27$, which is not significant. Thus, we conclude that the two groups of springs did not differ in their load characteristics. This comparison is important if load is assumed to affect pressure. Directly comparing pressures would not be meaningful if the spring loads differed appreciably.

(c) A scatter plot of the two pressure measurements appears in Figure 12.62c. Generally, the old spring results in higher pressure readings. Using Procedure 12.5, comparison of the pressure measurements can be made:

	Subassembly with Old spring	Subassembly with New spring	Difference
Mean	29.78	29.07	−.71
SD	1.17	1.07	.84

$T = (\sqrt{24} \times .71)/.84 = 4.14$

Thus, we conclude that the new spring results in a downshift of .71 psi. The practical importance of this shift must be evaluated.

12.14 (a) The prediction line factors are as follows:

$$b = \frac{.99 \times 3.28}{7.36} = .44$$

$$a = -.04 - (.44 \times 13) = -5.76$$

The prediction line is

Size $= -5.76 + .44$ hour

Figure 12.63 shows the tool wear prediction line.

(b) From the equation, the tool wears at a rate of 0.44 units/hour. Thus, it would require $20/.44 = 45$ hours to span the 20 unit specification width. This calculation assumes a constant wear rate. Since the actual interval in which data is observed is about −6 to +5, prediction beyond this interval is not justified.

(c) The long tool life gives a number of tool change options. It seems unnecessary to produce parts anywhere near the specification limits. An 8 hour tool change interval would span $8 \times .44 = 3.5$ units, or only about 18% of the specification width.

12.15 (a) In the scatter plot in Figure 12.64 it appears that the CMM gage reads consistently higher than the electronic gage.

(b) Using Procedure 12.5,

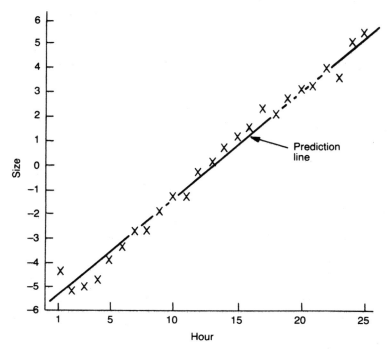

Figure 12.63   Tool wear.

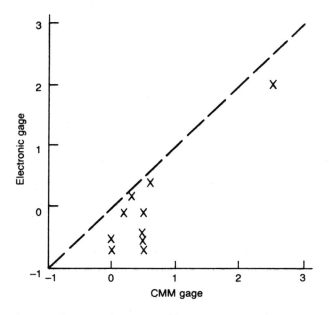

Figure 12.64   Scatter plot to evaluate gage bias.

$$D = -.55$$

$$s_p = .397$$

$$T = \frac{\sqrt{10} \times .55}{.397} = 4.38$$

which is greater than the tabled value of 2.26 so we conclude that a significant bias exists. The $-.55$ bias relates to the specification width by

$$\% \text{ accuracy} = 100 \frac{\text{bias}}{\text{tolerance}}$$

$$= 100 \frac{.55}{1.4}$$

$$= 39\%$$

Clearly, a bias of 39% of the specification width is unacceptable. The gage masters should be checked along with any gage elements relevant to bias.